Anthology of Biosafety: VIII

Evolving Issues in Containment

Other volumes of ***Anthology of Biosafety***, edited by Jonathan Y. Richmond, PhD, available from the American Biological Safety Association:

I. Perspectives on Laboratory Design (1999)

II. Facility Design Considerations (2000)

III. Application of Principles (2000)

IV. Issues in Public Health (2001)

V. BSL-4 Laboratories (2002)

VI. Arthropod Borne Diseases (2003)

VII. Biosafety Level 3

Anthology of Biosafety: VIII

Evolving Issues in Containment

Jonathan Y. Richmond, PhD, Editor
Jairo Betancourt, Associate Editor

American Biological Safety Association
1202 Allanson Road
Mundelein, IL 60060-3808
847-949-1517
Fax: 847-566-4580

AMERICAN BIOLOGICAL SAFETY ASSOCIATION
1202 Allanson Road
Mundelein, IL 60060-3808
847-949-1517
Fax: 847-566-4580

Edward J. Stygar, Jr.
Executive Director

Prepared under the auspices of the Publication Committee of the American Biological Safety Association.

Library of Congress Control Number: 99-60507
ISBN: 1-882147-69-3

Printed and Bound in the United States of America
Production Editor: Karen D. Savage
Cover Design: Richard Green
Printing: TK Graphics, Barrington, IL
Font: Arial

Anthology of Biosafety, VIII
Evolving Issues in Containment

Table of Contents

Preface

The need for all microbiological laboratories (research, clinical, biotech, pharmaceutical, etc.) to have and embrace a robust “biosafety culture” has never been greater. This is to say that all persons associated with the laboratories must not only understand the fundamental tenents of biosafety, but also the practices and procedures must be incorporated so successfully that they become the normal way of life. This is particularly true when working in and around containment facilities.

As more and more BSL-3, BSL-3 enhanced, BSL-4, ABSL-3, ABSL-3-Ag, and ABSL-4 facilities are constructed globally, there is growing concern that there are insufficient numbers of trained and experienced laboratorians, animal care personnel, and engineers to populate them. Certainly in North America and Europe we have begun to see an increased number of excellent training opportunities, including some that incorporate hands-on wet lab experience, and even some that focus on the appropriate use of personal protective equipment.

This type of training is essential to ensure proficiency in how to safely perform work with all high-consequence pathogens, including Select Agents. But training *per se* is only a beginning - experience comes after months or years of successfully performing such work. In addition to the laboratorians, three other professional groups need appropriate training and experience.

Laboratory administrators and managers (and others) are responsible for helping to create the broad policy and supportive infrastructure to establish the biosafety culture. They provide the hard resources (money, facility) as well as the soft resources (time, personnel, encouragement) from the top down. They need to be knowledgeable of the risks and be insistent upon appropriate risk management. They need to appreciate the consequences of breaches in containment, the need for appropriate biosecurity, and the need for routine auditing.

The facility engineers who are asked to maintain these containment laboratories on a daily basis need a practical understanding of biosafety. The systems that ensure tertiary containment are not particularly difficult to maintain; it is the rigorous consistent synergistic interaction, without interruption, that is crucial. There is also a need for open communication and teamwork (within the various trades, with the laboratorians, and with biosafety personnel) to build the necessary trustful working environment. Currently there are only a finite number of experienced containment engineers and only a dearth of training programs specifically directed towards these individuals.

Also key to the development and maintenance of the biosafety culture is the biosafety officer. The American Biological Safety Association and a few independent groups offer significant coursework for aspiring and practicing biosafety professionals. However, there is still the need for experience, particularly in providing service to the containment laboratories. There are currently only a relatively few fully-qualified BSL-4 biosafety personnel, a few more with experience at BSL-3. But there are a number of new labs in the planning stage or under construction. The NIH recently launched a program to provide advanced training / experience which should prove to be helpful.

The global needs are even greater, particularly in developing nations. The newly published WHO *Laboratory Biosafety Manual*, 3rd edition, also encourages the creation of an appropriate biosafety culture. Recognizing that there are relatively few available laboratory safety training courses, the WHO Biosafety Programme (BP) is working towards offering a biosafety Train-the-Trainer program. When smallpox virus was eradicated, the World Health Assembly charged the BP with performing routine audits of the two remaining maximum containment laboratories possessing the virus. It is expected that when wild poliovirus is eradicated, a similar charge will be given the BP to audit all laboratories and polio vaccine production facilities. A large number of qualified biosafety auditors will be needed. Similarly, it seems prudent that all WHO Collaborating Centers working with high consequence pathogens should be audited to assure continued compliance with the WHO biosafety standards, adding to the need for more biosafety professionals.

In this era of heightened concern regarding who has access to, who works with, and who transports these agents, it certainly seems prudent that there be a global network of competent biosafety professionals who truly understand the issues and are known to one another (at least by their recognized credentials). As we move forward, therefore, with the creation of many more high- and maximum- containment laboratories, we must work together to make sure all such facilities grow an appropriate biosafety culture.

- Jonathan
Southport, NC

Forward

The microbiological world has no natural boundaries, with new agents being discovered and rediscovered with alarming regularity. In addition, the perceived threats from the deliberate misuse of agents and the introduction of biotechnology into the natural world through genetic manipulation, have amplified the concerns of the public at large. In parallel, the development of new technologies have extended our capabilities to study the molecular structure of microbes, host responses and transmission dynamics, all aimed at achieving control strategies through the development of antimicrobials and vaccines.

Outbreaks of West Nile in the US, plague in India, Ebola virus in Zaire, Marburg in Angola, and SARS in Asia are examples of high profile infections that have raised public and governmental awareness of our vulnerability. The long term problems associated with the world-wide HIV epidemic, rising numbers of hepatitis B and C cases, and the unknown threats associated with variant Creutzfeldt-Jakob disease (vCJD), continue to fuel public fears. All these incidences have heightened the debate about how microbiological risks are identified and quantified, as well as how the principles of risk-based approaches should influence policy-making and operational practice. Most importantly there is a need to ensure that current and future microbiological hazards are managed against a background of sound scientific knowledge of risk.

Thus, increasing demands are also being placed on national and regulatory authorities to ensure that such pathogens are handled in a safe and secure manner. The resultant biosafety information included in documents such as the *Biosafety in Microbiology and Biomedical Laboratories* (US); *Laboratory Biosafety Guidelines* (Canada); *Categorisation of Biological Agents according to Hazard and Categories of Containment* (UK); and the *Laboratory Biosafety Manual* (WHO) provide the 'best practice' guidance to the scientific community.

Despite the greater awareness of biosafety and biocontainment practices, laboratories remain a source of infection and even mortality among laboratory workers. Incidents of secondary transmission of disease to the community caused by possible contamination of the environment or personnel have also been identified, although they are very few in reality. Caution must also be exercised when one considers that there is a steady increase in both the number of laboratories handling pathogens and in the number of scientists wishing to exchange new or exotic strains across the globe for further study.

The challenge for the scientific community is to continue with the development of biological containment systems, which balance the needs between scientific enquiry, and the restrictions associate with protecting laboratory workers and the community within which they work. Over the past two decades considerable energy has been directed at bringing together the skills of many disciplines to design, build, maintain and operate a variety of high containment facilities. Biomedical research and diagnostic facilities have become complex, requiring the interdisciplinary talents of designers, architects, mechanical engineers, production and electrical engineers, biosafety professionals and experienced scientists.

This Anthology reflects on current biosafety and biocontainment principles and practices. The collection of articles enables the reader to gain an insight into a variety of novel solutions applied to containing infectious substances in different working environments, enabling research and development to continue safely. Diverse contributions that address the specialised high containment issues surrounding the design, construction and operational needs of large scale production facilities requiring Good Manufacturing Practice, animal rooms, controlling TSE and the deployment of state-of-the-art mobile facilities to work at the site of disease outbreaks are contained herein. Articles concerning the development of biodefense infrastructures illustrate new challenges to the design and operation of biocontainment facilities. In contrast, the construction and operation of biosafety level 3 laboratories within new settings, admirably illustrate the issues required in less developed biosafety environments.

Current expansion of high containment laboratories throughout the world requires international cooperation and consensus to be established in facility design, commissioning, operation and maintenance, in accordance with our understanding of the infectious disease processes. It must be recognised that the operational integrity of these facilities will be dependent upon the establishment of good working relationships between scientists, specialist engineers and biosafety professionals. This will ensure that the primary missions of the laboratory (the research, developmental and detection programmes) can be conducted efficiently, safely and securely.

This Anthology reminds us that despite the tools being put in place, all involved in the establishment of high containment facilities need the knowledge and skill to enable their safe use.

Graham Lloyd, PhD
Special Pathogens Reference Unit, Head
Novel and Dangerous Pathogens
Centre Emergency, Preparedness and Response

Author Contact Information

Jairo Betancourt
Office of Environmental Health and Safety (R23)
University of Miami
P.O. Box 016960
Miami, FL 33101
JBetancourt@med.miami.edu

Jonathan Crane
Principal
CUH2A
400 Colony Square, Suite 600
1202 Peachtree Street, NE
Atlanta, GA 30361
www.CUH2A.com

Ellen J. Elliott
Baltimore Research and Education Foundation, Inc.
and
Research Service, VA Medical Health Care System
10 N. Green Street
Baltimore, MD 21201-1595
and
Department of Neurology
University of Maryland School of Medicine
22 South Green Street
Baltimore, MD 21202-1595
eelliott@comcast.net

Luiz Fernando Nunes de Azeredo
SQN 404 Bioco O Apto 303
Asa Norte
CEP 7845-150
Brasilia – DF
Brazil
Luiz.azeredo@funasa.gov.br

Leslie Gartner, AIA
Principal
Smith Carter
1123 Zonoiite Road, Suite 25
Atlanta, GA 30306
Lgartner@smithcarter.com

Vibeke Halkjær-Knudsen
Director Viral Vaccine Department
Statens Serum Institute
Artillerivej 5
DK-2300S
vhk@ssi.dk

Charles E. Henry
Technical Director
Edgewood Chemical Biological Center
ATTN: AMSRD-ECB-RT-AM
5183 Blackhawk Road
Aberdeen Proving Ground, MD 21010-5424
charles.henry@us.army.mil

Monica J. Heyl
Technical Director
Edgewood Chemical Biological Center
ATTN: AMSRD-ECB-RT-AM
5183 Blackhawk Road
Aberdeen Proving Ground, MD 21010-5424
monica.heyl@usarmy.mil

Chris Kiley, PE
Hemisphere Engineering
1123 Zonoiite Road, Suite 25
Atlanta, GA 30306
ckiley@hemisphere-eng.com

Claudia MacAuley
Baltimore Research and Education Foundation, Inc.
and
Research Service, VA Medical Health Care System
10 N. Green Street
Baltimore, MD 21201-1595
cmacaule@umaryland.edu

Nicoletta Prevasini, PhD
Project Leader, Biosafety
Department of Communicable Disease Surveillance
and Response
World Health Organization
AV. Appia 29
CG-1211 Geneva, 27
prevasinin@who.int

Edward H. Rau, RS, MS
Captain, U.S. Public Health Officer
Environmental Health Officer
Environmental Protection Division
Office of Research Facilities Development and Operation
DHHS National Institutes of Health
13 South Drive, Room 2W64
Bethesda, MD 20892-5746
raue@mail.nih.gov

Dennis J. Reutter, PhD
Technical Director
Edgewood Chemical Biological Center
ATTN: AMSRD-ECB-RT-AM
5183 Blackhawk Road
Aberdeen Proving Ground, MD 21010-5424
dennis.reutter@usarmy.mil

Jonathan Y. Richmond, PhD, RBSP
Jonathan Richmond and Associates
927 East Leonard Street
Southport, NC 28461
jonathanrichmond@bsafe.us

Deanna Robbins
Malcolm Randall VA Medical Center
1601 SW Archer Road
Gainesville, FL 32608-1197
deanna.robbins@va.med.gov

Robert G. Rohwer
Baltimore Research and Education Foundation, Inc.
and
Research Service, VA Medical Health Care System
10 N. Green Street
Baltimore, MD 21201-1595
and
Department of Neurology
University of Maryland School of Medicine
22 South Green Street
Baltimore, MD 21202-1595
rrohwer@umaryland.edu

Chapter 1

Toward Global Biosafety Harmonization

Jonathan Y. Richmond, PhD, RBP and Nicoletta Previsani, PhD

Introduction

During the spring of 2004 JYR had the unique opportunity of working in Geneva at the World Health Organization (WHO), assisting in the Biosafety programme (BP) with NP. The vision that is presented here evolved during that time, but it should not be taken as having been endorsed by WHO.

Since the early 1980's the WHO *Laboratory Biosafety Manual* (LBM) has been made available in many languages to provide a framework for microbiological and clinical laboratories, particularly in developing nations, to build good laboratories and practice safe procedures. The third edition of the LBM (LBM3) has recently been published (WHO 2004a). As with other, national biosafety manuals, (e.g., CDC/NIH 1999; MOH, Canada 2004) four levels of facility design and work practices are described. They provide a basis for working with microbial agents posing escalating risks to laboratorians, their coworkers, and the environment.

Biosafety level 1 (BSL-1) is reserved for work with agents posing low risks to the laboratorians and the surrounding community. BSL-2 is for agents posing minimal risk to the workers and low risk to the community, as these are the agents normally found circulating in the community. In reality BSL-1 and BSL-2 laboratories have quite similar design features and embrace rather similar work practices and procedures.

Agents with a high risk of spread through aerosol exposure are assigned to BSL-3. These laboratories are uniquely designed and operated to keep aerosolized microorganisms and their toxins within the confines of the immediate work area, relying on various levels of primary and secondary biocontainment to achieve that result. BSL-4 is reserved for work with high consequence pathogens for which there are no available treatment measures, including vaccines or antivirals. Accomplishing containment at BSL-4 relies on keeping the agents away from the worker, either by manipulating the microorganisms within Class III biological safety cabinets or by having the laboratorian work in a positive-pressure enclosed suit.

Biosafety Challenges

There is growing interest among public health officials for increasing awareness in and support for national biosafety programs. These programs would be of tremendous benefit in ensuring a consistency in occupational health and safety measures for the laboratorians who work daily with a variety of microorganisms in clinical, research and biotechnology activities. In the past few years there have been several notable laboratory-acquired infections reported that resulted from inappropriate biosafety practices (Harding and Byers, 2000; WHO, 2003; Lin et al, 2004; Rusnak et al, 2004; *Boston Globe*, 2005). Increased biosafety programs would be a major step forward in helping laboratorians work safely.

One of the exciting WHO initiatives currently underway is the global eradication of polio. As this campaign nears its conclusion, there has been considerable activity to plan biosafety guidelines to be adopted for laboratories and vaccine manufacturing facilities once wild polio virus is no longer circulating in our communities. Wild polio virus was not only found ubiquitously in all countries, it was a model virus used in many teaching, diagnostic and research laboratories. As the result of an active campaign to identify all laboratories possessing this virus and to encourage them to destroy or consolidate stocks of wild poliovirus, it is anticipated that soon there will only be a modest number of laboratories left who choose to continue to work with this virus.

In a world of diminishing immunity to polioviruses, the result of a release of wild polio virus could be catastrophic. To ensure that these remaining laboratories and vaccine production facilities follow the appropriate biosafety guidelines (not yet finalized by WHO), it is anticipated that there will have to be an audit / inspection program administered through the Biosafety programme office. The BP was charged with a similar responsibility following the successful campaign to eradicate smallpox virus. One could anticipate that as other infectious diseases are eradicated, the number of laboratories brought under such a program will increase further.

Scattered throughout the world are a fair number of WHO Collaborating Centres, many being laboratories engaged in agent-specific surveillance and research activities that are critical to global public health. Certainly it would be prudent and an affirmation of support for a robust biological safety program for all these Collaborating Center laboratories to meet the expectations provided by the WHO LBM. Perhaps they, too, should fall under a similar audit / inspection activity. Unfortunately there are currently insufficient resources (funding or personnel) within the BP for such activities to become a reality.

Biosafety Professionals

We live in an era of heightened awareness regarding the potential for additional bioterrorist events. We also recognize that emerging and reemerging infectious diseases pose additional threats to our collective way of life. Either or both of these events will challenge our global public health capacity. We need to improve our ability to communicate effectively and efficiently among our many countries during times of microbiological emergencies. It is critical to start educating biosafety professionals to become familiar with safe field working conditions, means for transporting infectious agent materials in international commerce, and strategies for creating safe environments for laboratorians. These professionals should be able to provide advice in a variety of technical microbiological matters in response to differing biosafety issues. The development of widely recognized biosafety officers is becoming a global need.

It is crucial that there actually be such individuals in each country, and even more important that these persons be knowledgeable of and conversant in the fundamentals associated with each of these specific areas. A strategy for moving in this direction has been undertaken recently by the International Biosafety Working Group. The IBWG, comprised of representatives of various biosafety entities, is working to create a mechanism to recognize persons meeting certain criteria as "biosafety professionals" (presented in Richmond 2004). The initial phase of this activity should be somewhat similar to the "Registered Biosafety Professional" program sponsored by the American Biological Safety Association (www.ABSA.org). The difficulty will be to achieve global consensus on such a registration program and global approval for registered candidates. The criteria will include certain formal education and work experience requirements, along with supporting documentation which may include specific professional recommendations. Experience in providing biosafety training, in conducting safety audits, in participating in laboratory design reviews, and involvement in safety committee activities are other important qualifications. WHO is currently working with countries to develop national biosafety professionals, possibly through the implementation of comprehensive education programmes for biosafety, at the Bachelor, Masters or PhD level.

Biosafety professionals could then become actively involved in biosafety-related initiatives. This might include active participation in national or regional microbiological or biosafety associations, serving as advocates for laboratory improvement, providing scientific-based advice during regulatory development, providing assistance to their national authorities during times of public health emergencies, and assisting in appropriate media-related activities. Biosafety professionals could also be active supporters of ensuring that appropriate national systems are available to expedite importation of infectious substances (e.g., materials needed for laboratory proficiency testing) or for their export (e.g., clinical specimens being sent to other laboratories for diagnostic evaluations).

Laboratory Reviews

There has been an increasing urgency to validate facility construction (typically called "commissioning"), continued facility operation (often during annual surveys) , and work performance (periodic safety audits). The commissioning process is performed by engineers, air-balancing technicians, and other specialists (often including the biosafety professional) who assure the owner of the facility that all systems that support the functionality of the laboratory meet the design parameters before actual work with infectious agents begins. The annual survey is often conducted by in-house facility engineers, technicians, and safety personnel. The survey is used to ensure that crucial support equipment (e.g., biological safety cabinets, autoclaves, air handling / filtration systems, etc) operate correctly. This is also an appropriate time to perform annual preventive maintenance activities.

Safety audits are generally performed by industrial hygienists and biosafety personnel, but also may be conducted by trained laboratory personnel. Audits can be conducted as a periodic check on the performance of each laboratorian, following an accident or near-miss, or as follow-on to annual training activities for validation of training effectiveness. Various techniques have been used in the conduct of safety audits; one type that is often found effective is the use of an audit check-list. Such documents can serve as a record of review and a basis for an on-the-spot teaching moment. They are also useful during subsequent audits to identify areas of improved performance.

Two sample audit forms are provided in Tables 1 and 2, developed from the 3rd edition of the WHO Laboratory Biosafety Manual, specifically based on chapters 3 (Basic laboratories - Biosafety Levels 1 and 2) and 4 (The containment laboratory - Biosafety Level 3). They are provided here to assist the biosafety professional who may be involved in these activities. More information on laboratory commissioning and laboratory certification is provided in chapters 7 (Guidelines for laboratory/facility commissioning) and 8 (Guidelines for laboratory/facility certification).

Action by the World Health

The 58th World Health Assembly (WHA) passed a significant resolution entitled "Enhancement of laboratory biosafety" (WHO, 2005). It considered that releases of microbiological agents and toxins may have global ramifications. It further acknowledged that the containment of microbiological agents and toxins in laboratories is critical to preventing outbreaks of emerging and re-emerging diseases. Clearly there is need for more biosafety professionals globally.

Summary

There is an ever-expanding need for more individuals to be recognized and certified for their proficiency in the science / art of biological safety. Our global community grows smaller and more intimately intertwined with serious issues related to the microorganisms that also inhabit our world. The natural evolution of modified microbes, the emergence of previously unknown ones, the potential for engineered agents that may pose a threat to humans, animals or plants - taken in the context of rapid air transport, crowded urban populations, encroachment into previously undisturbed environments - create the need for more biosafety professionals. This is both an exciting and challenging opportunity.

References

Boston Globe, (2004). US agency fines BU $8,100 in safety. Available at: http://www.boston.com/news/education/higher/articles/2005/05/10/us_agency_fines_bu_8100_in_safety_offenses/.

CDC/NIH., (1999). Biosafety in Microbiological and Biomedical Laboratories, 4th ed. US Department of Health and Human Services, Washington, DC: US Government Printing Office.

Harding, A.L., & Byers, K.B., (2000). Epidemiology of laboratory-associated infections. In Fleming, D.O. & Hunt, D.L. (ed). Biological Safety: Principles and Practices. 3rd Edition, ASM Press.

Linn, PL, Kurup, A, Gopalakrishna, G, Chan, RP, Wong, CW, Ng, LC, et al, (2004). Laboratory-acquired severe acute respiratory disease syndrome. *New England Journal of Medicine* 350,1740-45.

MOH (Ministry of Health), Canada, (2004). Laboratory Biosafety Guidelines, 3rd ed. Minister of Health, Population and Public Health Branch, Center for Emergency Preparedness and Response, Ottawa, Ontario.

Richmond, JY., (2004). Lessons Learned from Recent SARS Lab-Acquired Infections (Abstract), Conference Program, 47th Annual Biological Safety Conference, San Antonio,TX, October.

Rusnak, J.M., Kortepeter, M.G., Hawley, R.J., Anderson, A.O., Boudreau, E., & Eitzen, E., (2004). Risk of Occupationally Acquired Illnesses from Biological Threat Agents in Unvaccinated Laboratory Workers. Biosecurity and Bioterrorism: Biodefense Strategy, Practice, and Science. Vol. 2, No. 4.

WHO (World Health Organization), (2003). Severe Acute Respiratory Syndrome (SARS) in Taiwan, Chona, 17 December 2003. Availabe at: http://www.who.int/csr/don/2003_12_17/en/

WHO (World Health Organization), (2004a). Laboratory Biosafety Manual, 3rd ed. Availabe at: www.who.int/csr/delibepidemics/WHO_CDS_CSR_LYO_2004_11/en/

WHO (World Health Organization), (2004b). China's latest SARS outbreak has been contained, but biosafety concerns remain. –Update 7, 18 May 2004, Available at: http://www.who.int/csr/don/2004_05_18a/en/

WHO (World Health Organization), (2005). Agenda Item 13.9: Enhancement of laboratory biosafety. Available at: www.who.int/gb/ebwha/pdf_files/WHA58_A58-62-en.pdf.

Table 1. A laboratory biosafety survey form based on the 3rd edition of the WHO's Laboratory Biosafety Manual.

Biosafety Levels 1 & 2

Location:____
Date:__________**Person In Charge of Laboratory:**____________________

Survey Items	Yes	No	N/A	Comments
***Access*:**				
Biohazard sign posted where necessary (e.g., BSL-2)				
Access restricted				
Laboratory doors closed (BSL-2)				
Children not allowed in laboratory				
Non-research animals in laboratory				
Personal Protection:				
Effective personal protective equipment (PPE) provided				
Proper laboratory clothing worn				
Proper gloves worn for procedures				
Hands washed After glove removal Before leaving laboratory				
Proper eye protection worn				
Personal protective equipment worn only in laboratory				
Proper shoes worn				
Eating, drinking, smoking, applying cosmetics, handling contact lenses observed				
Human food or drinks not stored in laboratory				
Laboratory clothing properly stored				
Procedures:				
Mouth pipetting prohibited				
Materials kept away from mouth				
Procedures minimize aerosols				
Needle usage limited				
Written spill clean-up procedures followed				

Spills and accidents reported and recorded; records maintained				
Contaminated liquids decontaminated				
Work surfaces appropriately decontaminated before and after work				
Infectious specimens transported out of the laboratory in containers following approved regulations				
Materials removed from the laboratory appropriately decontaminated				
Laboratory working area:				
Area neat and clean				
Work surfaces properly decontaminated				
Contaminated materials decontaminated before disposal, cleaning or reuse				
Transport regulations followed				
Opened windows protected with arthropod-proof screens				
Biosafety Management:				
Biosafety manual developed, read, and followed				
Regular biosafety training provided				
Standard Operating Procedures (SOPs) and manual available in laboratory				
Arthropod and rodent control programme in use				
Laboratory design and facilities:				
General maintenance available				
Annual (regular) certification completed				
Design features:				
Ample space provided for safe work				
Walls, floors, ceilings and work surfaces easy to clean				
Floors slip-resistant				
Bench tops appropriate for laboratory work				
Adequate illumination available				
Furniture sturdy, cleanable				

Adequate storage spaces available and appropriately used				
Proper chemical, gas, and isotope storage				
Proper storage of street clothing is outside laboratory				
Hand-wash basins available near door				
Running water always available				
Doors have window panes, are fire resistant and self-closing				
Proper meal and rest areas outside laboratory				
Shower and eyewash available				
Fire fighting system available				
First aid materials available and not out-dated				
Reliable, adequate electrical supply				
Emergency generator available, tested				
Appropriate physical security in place				
Biosafety cabinet (BSC):				
Properly located in room				
Annually certified				
Personnel trained and proper procedures followed				
Open flames not used in cabinet				
Clean and orderly				
Used when infectious materials are handled				
If UV light available, used properly and checked regularly for efficacy				
Essential biosafety equipment and procedures:				
Pipetting devices available and used				
Screw-capped tubes, bottles used				
Plastic ware used (instead of glass) when possible				
Autoclave used, checked regularly				
Centrifuge tubes prepared, opened in BSC				
Vacuum lines protected with filters or disinfectant traps				
Waste containers for "sharps" used				
Adequate chemical disinfectants used				

Biomedical waste disposal appropriate				
Microwave ovens clearly labeled: "Not for food preparation. Laboratory use only"				
Health and medical surveillance:				
Occupational medical program available, including • Pre-employment health checks • Records of accidents, illnesses maintained • Immunizations offered relevant to the work of the laboratory • All laboratory accidents recorded • For BS-2, a targeted occupational health check performed • For BL-2, records of absences and illnesses maintained				
Appropriate medical services available				
Programme monitors for laboratory acquired infections				
Highly susceptible individuals excluded from high hazardous work area				
Women of child-bearing age provided with agent-specific risk counseling				
Training:				
All training documented				
New employees trained before they work				
Training provided for new procedures				
Laboratory supervisor ensures documented training includes information about risks from • Inhalation • Ingestion • Percutaneous exposures • Animal bites and scratches • Handling blood and pathological materials • Decontamination, infectious waste disposal				

Waste handling:				
Proper procedures used for decontamination				
Proper waste collection and removal procedures followed				
Waste handlers protected				
Sharps management practiced				
Decontamination:				
Proper procedures for autoclaves followed				
Decontamination methods appropriate for agent(s) and equipment in use				

Person conducting survey:____________________________________

Notes and Comments:

Table 2. A laboratory biosafety survey form based on the 3rd edition of the WHO's Laboratory Biosafety Manual.

Biosafety Level 3

Location:____
Date:________________ **Person in Charge of Lab:**________________________

Note: This survey is done <u>in addition to</u> the Biosafety Level 1 & 2 survey.

Survey Items	Yes	No	N/A	Comments
Code of Practice:				
Door posted with entry restrictions and supervisor contact information				
Proper protective lab clothing worn				
Primary containment device used for open manipulations of potentially infectious materials				
Respiratory equipment used if required				
Security procedures are in place				
Laboratory design and facilities:				
Laboratory is isolated from general traffic flow				
Double door access to laboratory • Doors are self-closing • Only one door can be opened at a time				
Change room/area available; shower optional				
Lab constructed for gaseous or vapor decontamination • Penetrations through walls, floors, ceilings sealed or sealable				
Windows closed, sealed, break-resistant				
Hand washing sink available near exit door is foot, elbow, or automatically operated				
Directional inward airflow maintained				
Visual air flow monitor available				
Room air exhausted without recirculation to other rooms				

Building exhaust air directed away from air intake				
HEPA filter installation allows for gaseous decontamination and testing				
Proper BSC installation and location within lab				
Autoclave available in laboratory, or proper procedures followed for safe removal of waste material to an autoclave				
Water supply protected with back-flow device				
Vacuum lines protected with liquid disinfectant traps and/or microbiological filters				
Laboratory commissioned				
Laboratory certified annually				
Laboratory equipment:				
Primary biocontainment devices used for all manipulations of potentially infectious materials				
Aerosol-containing centrifuge equipment available and used				
Personal Protective Equipment:				
Solid-front gowns, coveralls, scrubs worn in lab				
Appropriate gloves are worn				
Double gloving when necessary				
Eye protection available and worn when necessary				
Respirators available and worn when necessary				
PPE worn only in lab work area				
Biological Safety Cabinet: Type I or II				
Annual certification current				
Personnel trained				
Appropriate work practices observed				
Location appropriate				
Open flame burners not used in BSC				
Volatile chemicals not used in BSC				

Practices and Procedures:				
Personnel follow biosafety manual and good microbiological practices				
Personnel informed of specific risks				
Appropriate decontamination procedures used for Surfaces BSC Reusable materials				
Sharps precautions are practiced				
Handwashing procedures followed				
De-gowning activities are appropriate				
Infectious wastes autoclaved or otherwise decontaminated before disposal				
Health and medical surveillance:				
Mandatory medical examination for all personnel • Detailed medical history • Occupationally-targeted physical exam				
Medical contact cards are used				
Fever watch protocol followed				
Save serum program (optional)				
Agent-specific immunizations offered				
Emergency medical treatment available				
Exposures to infectious agents documented				

Person conducting the survey:______________________________

Notes and comments:

Chapter 2

Emerging Challenges in Containment: Development of a Biodefense Infrastructure

Jonathan Crane, AIA

Introduction

The terrorism of September 11, 2001 and the subsequent anthrax events in October of that year have focused efforts on the development of medical countermeasures to protect the civilian population against the use of bioweapons. In addition, the possible terrorist use of plant or animal disease agents against the agricultural sector of the United States poses a significant threat to our economic well-being.

The development of the laboratory infrastructure to begin developing a response to these threats has quickly moved from planning to reality. The requirements of these laboratories pose new challenges to the design and operation of biocontainment facilities. This chapter will identify how the national strategy is shaping up for the development of countermeasures and focus on some of the unique new challenges in containment of the laboratories designed to support this effort. It should be noted that much of the efforts for biodefense are indistinguishable from the efforts required for developing responses to naturally emerging infectious diseases; however, some laboratory requirements may differ.

The Case for a Biodefense Infrastructure: Is There a Threat?

There has been questioning in the scientific, military, and political communities as to the level of threat from bioterrorism. The difficulties of successfully using biological agent's at large scale in warfare have contributed to a sense that they would be ineffective as terrorist agents as well. It is clear, however, from the anthrax letters in October 2001 that even a small-scale terrorist use of a biological agent can cause disruption disproportionate to the actual biological threat. In addition, the legitimate question exists as to whether funding research on the potential threat of biological terrorism takes away funding from current public health threats from infectious disease such as SARS, AIDS, and multiple drug-restraint strains of disease (Sidel, 2002).

The realization that the threat represents a significant threat to health and economic well-being is now well documented and is a matter of public policy. The United States Senate has found that "the threat of biological and chemical weapons is real." It has enumerated current threat agents including those developed by the former Soviet Union in its biological weapons program. In addition, the Senate found that "the possibility exists that terrorists or others will use biotechnology techniques to enhance the lethality of a biological agent. According to the Defense Science Board, "Motivated researchers using advanced genetics techniques can engineer pathogens with unnatural characteristics that enhance their offensive properties by altering such characteristics as stability, dissemination properties, host range, contagiousness, resistance to drugs and vaccines, and persistence in the environment, among others" (Lieberman, 2003). In this vein, the scientific and intelligence communities are beginning to realize that potential future threats are significantly greater than first thought (CIA, 2003). These threats, if realized, have the potential to significantly disrupt not only public health but also life sciences research and development.

The Inevitable Dissemination of Technology

As evidenced by the proliferation of nuclear weapons and technology over the last fifty years, it is difficult if not impossible to keep technology from disseminating. This will be particularly true with technology related to biological weapons development as much of the basic knowledge required for the development and production of novel weapons has been published and the technology to allow development is dual use. Without countermeasures to adequately protect the public against the potential threats, the only option available to governments will be to develop nonproliferation controls.

Unlike the controls over materials required for nuclear threats which, due to the rarity of the fissionable material required for devices, have little impact on the general public, the controls that would be required to slow the proliferation of biological threats will have a far-reaching effect on many aspects of our society. The scientific and biotechnology community that has, through self-governing ethics, been largely free to pursue paths to knowledge and technology development will have restrictions placed on the development of knowledge or technology for potential threats where countermeasures do not exist. This has the potential for a large disruption in these communities because many things that have potential benefits with appropriate use pose potential threats with misuse.

A compelling argument can be made that the threats are too diverse and unpredictable to develop countermeasures that will adequately protect society from the full range of threats (Cordesman, 2001). This may well be true; however, without novel means of protection against these novel threats, the one certainty is that our society will undergo significant disruption and change. Our only real option is to begin the process of understanding the nature of the threats, prioritizing the current and future threats, and begin the process to develop effective countermeasures.

The Difficulties of Countermeasure Development

There has been an ongoing effort to develop vaccines and therapeutics for protection of our troops against biological warfare since World War II. While significant research has been accomplished, little of the efforts have translated into approved products to protect our military (Cohen, 2001). The effort to develop countermeasures becomes increasingly complicated when applied to the civilian population. Issues of age, underlying health problems, and immune system suppression and deficiencies make vaccines less effective as universal countermeasures.

Challenges in countermeasure development for infectious agents include lack of understanding of the organisms of interest, lack of understanding of the diseases caused by the agents, lack of a patient population, and lack of animal models to study the diseases. The research to gain this knowledge will be hampered by the difficulty of working within containment. The normal productivity encountered by biotechnology and pharmaceutical companies will be difficult to achieve due to the above issues increasing the length of time and quantity of personnel required to provide the products to protect us. Security requirements (CDC, 2002) will also impact productivity and the availability of scientific personnel.

The greatest challenge will be to overcome the scale of the infrastructure shortage of both facilities and scientific personnel for research and development of the countermeasures that may be required. It has been estimated that 100 to 200 new products will be required to adequately protect the civilian population against the known threats and potentially many more against novel threats yet to be identified. The pharmaceutical industry utilizes approximately 4,000 research and development employees per product. The currently planned biodefense infrastructure in the aggregate will house approximately 2,500 employees. It will be difficult to fully staff this initial infrastructure; however, the development of countermeasures is the only option for securing our future against the threats.

It has been said that a great journey starts with a single step. The first steps are being taken to provide the infrastructure, both physical and scientific. Much of this is occurring in the government and academic sectors of the scientific infrastructure. The next steps must be to develop the infrastructure in the government, academic, and private sectors to adequately meet the need.

Facility Requirements for Countermeasure Development

The facilities for biodefense research and development are of unprecedented scope, scale, and complexity compared to facilities previously operated for biocontainment. The process for development of vaccines and therapeutics begins with an understanding of the biology of the organisms in question and the pathogenesis of the diseases they cause. Much of this work includes biocontainment laboratories. Biocontainment laboratories at either BSL3 or BSL4, and will be required for much of the research and development process for growth of the organisms in question and extraction of their nucleic acids and proteins. The biocontainment level will be defined by the risk assessment of the agent and the proposed work per the CDC/NIH *Biosafety in Microbiological and Biomedical Laboratories* (Richmond and McKinney, 1999). In addition, advanced research equipment such as confocal microscopes, cell sorters, and cell counters are being housed in containment to be used with viable infectious organisms. Basic molecular biology laboratories are used in conjunction with the containment laboratories to work with inactivated agents, nucleic acids, and proteins. Containment animal facilities are used to research the transmission and progression of disease *in vivo.* Animal facilities are also required for the development and validation of animal models to mimic the progression of the disease in humans.

A relatively new development in contained infectious disease research is the use of advanced imaging technologies such as magnetic resonance (MR), computerized tomography (CT), and positron emission tomography (PET) to evaluate live infected animal models for structural, pathological, and metabolic information on the progression of disease. The introduction of this equipment poses new challenges in containment due to the need for movement of infected animals from contained animal housing to the equipment also located in containment while maintaining biosafety protocols and minimizing the potential for cross contamination.

Development of vaccines and therapeutics require similar laboratories and animal facilities to those described above. In addition, noncontainment laboratories are required to develop the drug candidates for *in vitro* and *in vivo* challenges against the agents. Low and high throughput screening facilities may be required in containment to check candidate libraries against the agents in question. Once candidate vaccines or therapeutics show promise, they are tested for efficacy against the agent in preclinical trials involving animal models. As many of these agents are infectious via the aerosol route of exposure, aerosol challenge facilities may be required to prove efficacy versus disease from this route of exposure.

Once a drug or vaccine has been accepted by the FDA as an investigational new drug (IND), the process continues with drug safety assessment in noncontained animal challenge facilities and human clinical trials similar to the process for all new drugs. A major difference in the development of drug for biodefense is the lack of a patient population for the efficacy testing of these drugs against humans. Due to this, the FDA has developed "animal rule trials" that allow efficacy testing of the drug or vaccine against two animal models that are sufficiently well characterized to mimic the effects of the drug on the disease in humans. For the agents to be studied to date such as anthrax, the guidelines developed by the FDA require significant animal quantities studied and evaluated for a considerable length of time (FDA, 2002). These studies will require a large amount of animal space in containment.

The manufacturing of vaccines for biodefense may require limited containment for live attenuated virus vaccines depending on the risk assessment of the growth of these organisms on a large scale. The manufacturing of therapeutics will be handled in normal facilities for the type of drugs to be produced.

Issues in Biosafety Management

The facilities described above will present new challenges in the biosafety management of the operations in the facilities. The scope of the larger facilities will be very comprehensive. As described above, facilities will include basic and containment laboratories and animal facilities as well as specialized contained facilities such as clinical labs, necropsy suites, surgical suites, and pathology labs. Specialized equipment will be used such as centrifuges for separation of cells and cellular components, lyophilizers for freeze-drying specimens, and flow cytometers for cell sorting and counting.

Cellular imaging technologies will be employed with viable organisms such as cryo-electron microscopy to look at molecular structure and advanced three-dimensional computerized microscopy to examine cellular structure and function. In addition, whole animal imaging will be done with a host of technologies new to containment, each with unique issues and characteristics. MR, SPECT, PET, CT, X-ray, and thermal imaging will require transfer of animals in containment. Scale-up facilities for both bacterial fermentation and mammalian cell virus culture may be required to produce vaccine candidates as well as cellular components for testing and evaluation of new drugs and vaccines. This broad scope of functions and complexity associated with these activities both inside and outside of containment, utilizing many different agents and containment protocols will require risk assessment, protocol development, and training on a continual basis.

The scale of the facilities will also be unique. To develop the required countermeasures, it will not be unusual to have 50 to 100 personnel needing training and monitoring in containment in a large biodefense research facility. Research scientists, technicians, veterinarians, animal care personnel, and maintenance personnel will all require differing protocols and procedures to perform their duties successfully. Training laboratories will be a major component of these facilities.

Roles of the Various Federal Agencies

The federal government has begun to respond in a coordinated effort to create the infrastructure for preparedness and response to terrorism from biological agents or their toxins. While all the responsibilities and interagency relationships have yet to be fully developed, general areas of responsibility that seem to be developing are as follows:

1) The National Institutes of Health (NIH) is responsible for research efforts on potential threats. They have categorized the current threat agents and have published a research agenda for the agents. The main institution within NIH for carrying out this agenda is the National Institute of Allergy and Infectious Diseases (NIAID). NIAID supports both intramural efforts in developing the science required with its own personnel and extramural efforts by the awarding of grants and contracts to academic and other research institutions. It has developed a strategic plan (NIH, 2002) and a research agenda (NIH, 2002) as well as programs such as the Centers of Excellence in Biodefense and the National and Regional Biocontainment Laboratories to develop infrastructure to support the research effort.
2) The Center for Disease Control and Prevention's role is one of preparedness and response. CDC, in conjunction with the State Public Health Laboratories and medical care providers, has developed the Laboratory Response Network. This network is designed to provide an early identification system in the event of a bioterrorism incident. CDC performs the public health diagnostic and epidemiologic functions for the federal government. They also identify and stockpile vaccines and therapeutics that may be required for quick response in the event of an incident.
3) The Department of Homeland Security (DHS) has the role of assessing the threats and providing a coordinating effort to the other federal agencies. It is also the lead agency for developing countermeasures against agroterrorism.
4) U.S. Department of Agriculture will coordinate research with the Department of Homeland Security to protect against plant and animal agents.
5) The Food and Drug Administration will be responsible for the approval and licensing of countermeasures.
6) The Department of Defense will continue to be responsible for the development of vaccines and therapeutics to protect the military.
7) Other agencies such as the Department of Justice and the Department of Energy have roles in supporting both the research and response efforts.

Role of Academic and Other Institutions

Academic institutions, both public and private, form the backbone of basic science research in the United States, particularly in the area of basic research. This basic research necessary to spur the development of vaccines and therapeutics for biodefense threats must rely heavily on the academic community. The availability of extramural grants from various federal agencies, particularly the NIH, is causing many institutions with expertise in infectious disease research to create the infrastructure to begin to develop a response. Many of the academic and corporate relationships and models for produce development that have been utilized for biotechnology products can be leveraged to create similar relationships for biodefense.

Role of the Private Sector

Past performance in the development of vaccines and therapeutics by the government makes it unlikely that the challenges can be overcome without the significant involvement of the private sector, particularly the pharmaceutical and biotechnology industries. Significant obstacles must be overcome to entice these industries to shift resources from current research and development efforts and apply new resources to development of countermeasures. Public policy debate needs to set a balance to ensure that the development of products for biodefense does not adversely affect the development of products for chronic and acute diseases that currently affect society. It is also important to ensure that resources are available to protect the public from naturally occurring infectious disease epidemics such as the flu and SARS.

Conclusion

The ability to provide countermeasures to possible bioterrorism is a complex issue with choices and implications that will affect many areas of our society. Although initial efforts are underway in the development of countermeasures, it is unclear at this time whether the current public policy debates and measures enacted will provide sufficient incentives for the private sector to significantly contribute to the research and development effort. Past efforts in conventional defenses have proven successful in developing an industry to respond to the challenge with the government being the major fund source for research and development programs and the major purchaser of the systems produced.

Current steps will move us forward in understanding the problem and defining the issues to be resolved. The true nature of the future threats will make it a problem that will be ongoing as novel threats are identified and countermeasures developed. While it is unlikely that we will apply the resources or resolve to stop all threats, we can act to limit the adverse consequences to society.

References

CDC (Centers for Disease Control and Prevention), Laboratory Security and Emergency Response Guidance for Laboratories Working with Select Agents, U.S. Department of Health and Human Services, *MMWR* 51 (RR-19); 1-8, December 6, 2002.

CIA (Central Intelligence Agency), (2003, November 3). Directorate of Intelligence, the Darker Bioweapons Future (Unclassified), OTI SF 2003-108

Cohen, J. and Eliot, M. Vaccines for Biodefense: A System in Distress, Science, Vol. 294, October 19, 2001

Cordesman, A.H., Biological Warfare and the "Buffy Paradigm", Center for Strategic and International Studies, Washington, DC, September 29, 2001

FDA (Food and Drug Administration), Center for Drug Evaluation and Research (CDER), FDA Guidelines Inhalational Anthrax (Post-Exposure) Developing Antimicrobial Drugs, Draft Guidance, March 2002

Lieberman, J., & Hatch, O. (2003, March 19), Biological, Chemical and Radiological Countermeasures Research Act, Senate Bill 666

NIH (National Institutes of Health), NIAID Strategic Plan for Biodefense Research, National Institute of Allergy and Infectious Diseases, February, 2002

NIH (National Institutes of Health), NIAID Biodefense Research Agenda for CDC Category A Agents, National Institute of Allergy and Infectious Diseases, February, 2002

Richmond, J.Y., & McKinney, R.W. Eds. Biosafety in Microbiological and Biomedical Laboratories, U. S. Department of Health and Human Services, Centers for Disease Control and Prevention and National Institutes of Health, 4th Edition, 1999

Sidel, V. W., Gould R. M. & Cohen, H. W. (2002). Bioterrorism preparedness: Cooption of public health? *Medicine and Global Survival*, 7, 2.

Chapter 3

Brazilian Experience in Implementation of Biosafety Level 3 Laboratories (BSL-3)

Luiz Fernando Azeredo

Introduction

Identification of a new hazardous virus in Brazil is not something uncommon; however, the country has yet to be equipped with Biosafety Level 3 laboratories (BSL-3) appropriate for the manipulation of such organisms. To resolve this situation, the federal government determined, through the Health Ministry, the implementation of the first Brazilian BSL-3 laboratories which are now in their final phase of construction.

To develop a program of requirements and physical programs for those laboratories, a multidisciplinary committee was formed that consisted of architects, engineers, biochemists, and biomedics. This committee visited the CDC, in the United States, and then covered the various reference institutes in Brazil, aiming to identify the ones that could support areas such as a biosafety level 3 and to assess how these laboratories should perform in order to be adapted to Brazilian technological and epidemiological settings.

This chapter analyzes the development of the program, from the requirements to the architecture and engineering projects of the first Brazilian biosafety level 3 laboratories. This analysis is based on the interaction of four fundamental parameters: the experience of "bench work" researchers, the technical knowledge of architects and engineers, specialized literature, and consultation to relevant norms concerning the subject. The author is an architect and a participant in the multidisciplinary committee and this analysis is a synthesis of the author's Masters Dissertation, held at University of Brasilia - UnB, under the advice of Prof. Dr. Jaime Gonçalves de Almeida.

Antecedents in Brazil

The concern with biosafety in Brazil first began with some public health occurrences that were not totally clarified, and served to show how unprepared Brazil was to deal with that kind of situation at the time, from the lack of technical training to the inexistence of appropriate installations.

In May 1995, the World Health Organization and the government of Zaire reported an outbreak of hemorrhagic fever caused by the Ebola virus that occurred in that country, in the Kikwit city and neighboring areas. The great repercussion of that news in the worldwide press, as well as the arrival of Brazilian journalists that went to Zaire, stimulated the Health Ministry - MS (Ministério da Saúde) to promote a meeting, dated May 22, with the presence of 17 Brazilian experts. From this meeting some recommendations were drawn, in which the highlights were:

1) The accreditation of Institute Evandro Chagas/MS (Belém, PA), Oswaldo Cruz Foundation - Fiocruz/MS (Rio de Janeiro, RJ), and Institute Adolfo Lutz/SES (São Paulo, SP), as reference laboratories, and the commitment from MS to provide these centers with resources to enable an inter-relationship with international reference centers.
2) Implementation of maximum security laboratories, destined to study viruses isolated in Brazil and which are classified as Biosafety Level 4 agents.

In the end of May 1995, three officials from the Brazilian Army presented with a respiratory syndrome associated with pneumonia, followed by the death of one official. They had been participating in a jungle survival course, near the road that connected Itacoatiara and Manaus. Due to the recent events in Zaire, there was a certain amount of panic at the time, mainly at Emílio Ribas Hospital in São Paulo, where the officials were hospitalized.

The institutes Evandro Chagas and Adolfo Lutz were summoned to urgently analyze the samples collected from the patients. However, it wasn't possible to establish a definite diagnosis with clinical, laboratorial or epidemiological basis (Fiocruz, 1997).

Due to this indefinite diagnosis, the Army and Health Ministries decided to constitute an official Brazilian commission, in charge of taking biological specimens to be analyzed in the United States. The commission also received the recommendation to initiate understandings with the Center for Disease Control and Prevention - CDC, from Atlanta and U.S. Army Medical Research Institute for Infectious Diseases - USAMRIID, in Maryland, for a possible visit to Brazilian institutes and technical supervision for the building of maximum security laboratories in Brazil.

The Sabiá Virus

The lack of appropriate installations in Brazil has already been responsible for widely divulged episodes with unknown substances. According to Fiocruz (2000), in January 1990, a new virus was isolated in the region of Cotia, São Paulo, from a fatal case of hemorrhagic fever. The name *Sabiá* came from the neighborhood in which the virus was detected. In the identification, done at Institute Adolfo Lutz/SP, no similarity to other viruses was found. The isolate was then sent to Institute Evandro Chagas, in Belém. There, during the characterization of the virus, a technician from the laboratory was infected, probably by aerosol, and was hospitalized in an Intensive Care Unit for 15 days. A sample of this virus was sent for analysis to an Epidemiology Department, in Connecticut, USA. The proceedings with the virus at that laboratory were suspended after some of its technicians were contaminated. Finally, it was recommended than an analysis in a BSL-4 laboratory at CDC, Atlanta is done. After the agent identification as a novel virus and because of the accidents during the many steps in its isolation in the USA and Brazil, all scientific activities regarding the Sabiá virus were halted in both countries. The genetic sequencing of Sabiá has already been performed abroad and presented at conferences. Even though it was initially isolated in Brazil, the knowledge of that virus was lost when its samples were sent to the U.S.

BSL-3 Laboratories in Brazil

The facts shown at the beginning of the chapter, which occurred in some countries and especially in Brazil, involved contamination by microorganisms of high risk to the community and the environment. They heightened the biosafety concerns of professionals in the scientific area and stimulated discussions with the government over the investments in maximum security laboratories in the country. In the year of 2002, the building of twelve Biosafety Level 3 laboratories was approved in Brazil, with funds from the World Bank.

To develop the program of requirements and projects for these laboratories, a committee was formed, called the national committee. It was comprised of architects and engineers with technical knowledge in potentially contaminated health buildings and containment areas, as well as researchers that worked directly with laboratory procedures involving BSL-3 agents, who knew the risks and the required schedule for such procedures. The participants on the committee are professionals from the National Foundation for Health (Fundação Nacional de Saúde) - MS, from the company hired to develop the projects and from specialists summoned to advice in specific areas.

The institutes that already carried studies with BSL-3 agents were the ones chosen for the implementation of the laboratories. The criteria used for the selection were experience with the agents, epidemiological relevance, and strategic geographical position.

The laboratories being constructed in Brazil should meet the long-standing need of the institutes and of the country. However, it will still be necessary to change habits in procedures and maintenance. Maintenance is one of the most expensive items in a BSL-3, and the monthly expense to keep it working is beyond the budget of some Brazilian institutes. All clothing (Figure 2) used to enter the containment area must be either discarded or autoclaved. The equipment is highly expensive and so is the air conditioning system maintenance. Therefore, it is essential that the Brazilian biosafety laboratories are administered competently when they begin their activities, so that they do not become a burden in the future and that they achieve their goal, which is to provide a safe environment and autonomy to Brazilian research institutes and preserve our biological diversity.

Figure 1. Brazil map showing the location of the institutes in which the laboratories are being built

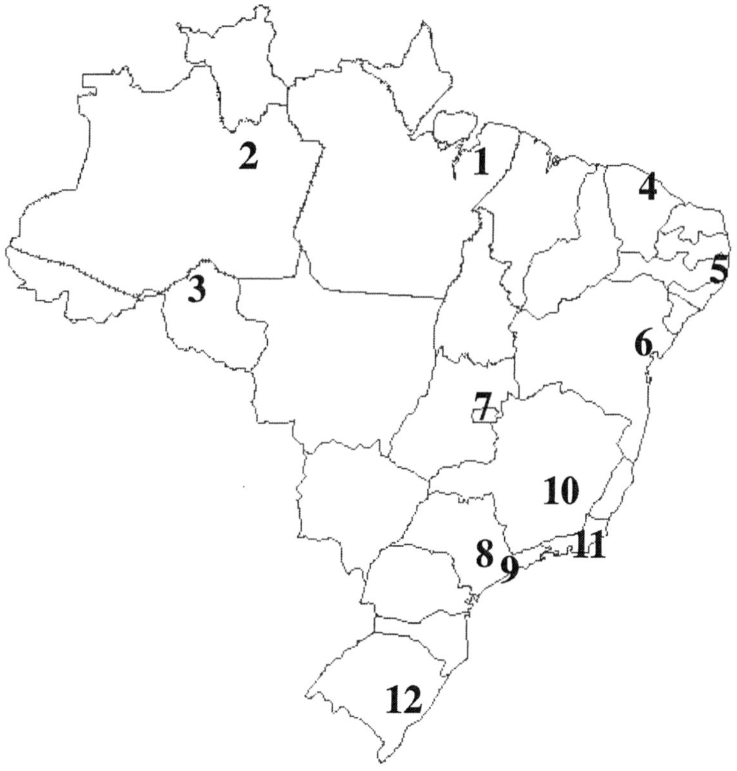

Footnote:

1. Belém/PA (Instituto Evandro Chagas/Secretaria de Vigilância em Saúde) – Arbovirus, Influenza Aviária e West Nile;
2. Manaus/AM (Fundação de Medicina Tropical/Secretaria Estadual de Saúde) – Arbovirus e Hantavirus;
3. Porto Velho/RO (Centro de Pesquisas e Medicina Tropical - SES) – Arbovirus;
4. Fortaleza/CE (Laboratório Central) – *Yersinia pestis* e *Mycobacterium tuberculosis;*
5. Recife/PE (Centro de Pesquisas Aggeu Magalhães) – *Yersinia pestis*, Hantavirus e *Mycobacterium tuberculosis*;
6. Salvador/BA (Centro de Pesquisas Gonçalo Muniz) – HIV e *Mycobacterium tuberculosis;*
7. Brasília/DF (Laboratório Central) – *Mycobacterium tuberculosis;*
8. São Paulo/SP (Instituto Pasteur) – Rhabdovirus;
9. São Paulo/SP (Instituto Adolpho Lutz) – *Mycobacterium tuberculosis;*
10. Belo Horizonte/MG (Instituto Otávio Magalhães/ Laboratório Central) – Arbovirus e Rickettsias
11. Rio de Janeiro/RJ (Departamento de Virologia do Instituto Oswaldo Cruz/Fiocruz) – HIV e *Bacillus anthracis;*
12. Porto Alegre/RS (Laboratório Central) – Rhabdovirus e Hantavirus.

Figure 2. Required PPE* for a BSL-3 laboratory

*PPE: Personal Protective Equipment Picture: L.F. Azeredo

Analysis of Component Elements of the Construction

The analysis of a BSL-3 laboratory involves components of architecture and engineering systems, denominated in analytical aspects. It is based on the fact that the experience acquired with planning and constructing this type of building is important to refine the existing norms. This vision is shared by the federal institutions and also by professionals, because of the pioneer characteristics these experiences have in Brazil.

In this sense, this analysis greatly relies on the acquired knowledge of the different professionals participating in this task, regarding project development, bench work research, norms and literature elaborated by the multidisciplinary teams. The main points taken into consideration were the particular characteristics of materials and systems versus the functional and technical requirements of these biosafety buildings.

Layout and Flow/Circulation

Some laboratories were projected to be built next to existing laboratory areas of the institute, while in other institutes there was no free area available for construction and the laboratories were projected to be implemented inside the existing buildings. This last case presented a higher level of difficulty, to provide a BSL-3 environment and locate the equipment in an already built and used technical floor, compared to the places in which it was possible to build an area specifically for level 3.

The layout of the analyzed laboratory was planned in such a way that the BSL-3 containment environment was preceded by a BSL-2 environment, restricting the access to BSL-3 area to the maximum. The benches will be 0.90 m tall and 0.60 m wide, considering the length of the equipment that they will support. The bench tops with sinks will be made of stainless steel, while the ones without sinks will be made of black colored epoxy. The benches will have a stainless steel tubular structure, with individual leveling devices (swivel levelers).There will be highchairs, appropriate to the bench height, without wheels or arms, and will be furbished with nonabsorbent and easy-cleaning synthetic material.

Some central benches will have narrow shelves above them, at about eye level. The other benches are located by the walls and won't have tall shelves, to facilitate cleaning and prevent dust and residue accumulation. To store smaller material, wheeled drawers or cupboards with shelves and doors are planned. They will be installed under determined benches. In this way, they can be moved around to where they are necessary, and can also free the space in case the researcher needs to perform procedures in which he/she has to sit.

Walls

The national committee considered PVC painting the most adequate wall finishing for the laboratory. The product has the necessary elasticity to prevent cracks and the desired resistance to bear constant cleaning with chemical products. Besides, it allows a smooth and impervious finishing and meets the recommendations and norms cited previously.

Among all mobile dividers analyzed by the national committee and researchers, the most adequate was made of aluminum frame coated by electrostatics painting, because it was lightweight, could be disassembled, had a smooth finishing for easy cleaning, and allowed for easy installation of glass or any rigid surface for closure. All divider walls will be inside the containment area, so sealant dividers won't be necessary, making it possible to use material easily found in the market with accessible price, as compared to stainless steel. This divider meets the requirements of NIH (2000), the norms of Anvisa (2002) and NBR 14050. It also allows the installation of electrical ducts that will be used to hold all electrical and analogical cables of the laboratory.

Flooring

The floor from the decontamination area specified by the technical committee was the ceramic tile with epoxy joint, due to its impermeability, resistance to disinfectants generally used for cleaning and washable properties. Even though this floor needs joint fillers, which could offer a restriction for its use, it is made of epoxy and meets the norms of Anvisa (2002). Anvisa requires that both the joint filler and the ceramic tiles present a water absorption index greater than 4%.

In the BSL-2, BSL-3 areas and the antechambers, the technical committee opted for highly resistant vinyl flooring. This floor is slip-resistant, is smooth and easy to clean, impermeable, and resistant to most chemical products and disinfectants used in the cleaning procedure. It will also be prolonged with a 10-cm-tall rounded baseboard with the least curvature possible (around ½ cm). Behind this curvature paste PVC will be injected to prevent occupation of free space by microorganisms or small insects.

Where the vinyl baseboard meets the wall, a vinyl filament will be used to allow an aligned and flat finishing between the two materials, facilitating cleaning and preventing dust accumulation. In this manner, NBR 14050 requirements are met for mineral aggregates and epoxy-based coatings of high performance, as well as Anvisa requirements (2002) regarding cleaning, impermeability, finishing of baseboard joint, resistance, and safety.

According to the national committee, an epoxy-coated floor was recommended for the showering area, because it is impermeable and slip-resistant. The national committee considered burned cement as the best option for the technical floor, because it is slip-resistant, and of low-cost and maintenance, besides meeting the requirements of NR 8 of Labor Ministry (Ministério do Trabalho) for minimum requirements for safety and comfort of workers in the building.

Ceiling

As a result of technical discussions with the researchers and the national committee, it was established that it would be necessary to use ceiling under the slab providing sufficient space for the passage of electro ducts, water system and air conditioning ducts. The space for those installations should be taken into account for the calculation of the laboratory headroom.

From all the options researched, the plaster panel ceiling was considered the most adequate by the technical committee and the researchers. The advantages of this material are stringency, imperviousness, resiliency for sealing wall corners, and easy replacement panels. It also allows the same coating used in the walls for a smooth finishing and easy cleaning. The committee specified acrylic painting for the ceiling. It complies with the requirements for cleaning, disinfecting, and resistance to chemical products used in the laboratory.

Door and Window Frames

The project committee defined a model and a standard measurement for the doors that give access to the BSL-2 and BSL-3 areas, according to the requirements cited previously and to the manufacturers' greatest measures.

The doors will be the side-hinged swinging type and will have two panels, one 0.80 m and the other 0.30 m, for the passage of large equipments. The other doors will have only one panel 0.80 m wide. The doors and windows will have aluminum frames with electrostatic painting and glass panels, with the exception of the doors in the antechambers and dark room, in which, instead of glass, compact formica will be used. The melaminic panel TS was the one which better complied with the requirements, due to its great mechanical resistance. It is structural, totally opaque, and 10 mm thick, close to the selected 8 mm thick glass. Only the windows in the BSL-2 area will be opened - the BSL-3 windows will be sealed and permanently closed.

Glass

The interlayer polymer film laminated glass was the one selected by the technical committee and researchers. From all the other glass analyzed, it was the one that fit the technical requirements, such as smooth surface, transparency, low heat absorption, and holding of glass fragments in case of breakage. There are options for thickness and laminate colors in the market. The one specified for the laboratory was 8 mm, composed of 3 mm uncolored float glass, 0.36 mm uncolored PVB film and 4 mm uncolored float. This combination of glass and uncolored film should prevent the absorption of a great part of infrared rays and, in case it breaks, will result in little or no glass fragments and will be safer for researchers. This glass can be found in nearly all regions and can be quickly replaced, if necessary.

Air Conditioning and Filtering

For air cooling, the technical committee and specialists determined that the best option for BSL-3 would be a ventilation system with variable air volume, which would maintain the temperature using the variation of air outflow. That would be possible with the installation of a thermostat inside the laboratory which would control the outflow and keep room temperature around 23°C ± 2°C. This is considered a comfortable temperature and will be adopted in accordance with NBR 6401 that establishes basic comfort parameters for air conditioning systems. This will enable researchers, who will be wearing thin and light special clothing, to spend the necessary time inside the laboratory without feeling cold.

A monitor will be installed in the laboratory entrance, before the containment area. It will show all basic information on the performance of the laboratory installations. Before the researcher enters the BSL-3, he/she will be able to know, among other things, if the air flow is correct, if the pressure is 40 Pa, if the exhaustion pipes are functioning properly, if the BSCs are turned on, and if the HEPA filters are inside the specifications. Visualization and system checking provide more safety to the researcher before his entrance in the containment area to begin a procedure.

The project committee, the air flow specialists, and the clinical engineers determined that the air humidity would be maintained around 50% ± 5%. This parameter is inside the margin established by NBR 6401, provides comfort for the user, and also allows good operation of electronic equipment.

Electrical Installation

The technical committee and the electrical engineers decided to specify two chillers that, together, will supply the air conditioning needs. The dimensioning was done in a way that, even if one of them stops working because of technical problems, the other will be capable of supplying the laboratories' demands in a low-consumption regimen, while the problem is fixed.

The distribution of electro ducts was planned by the electrical engineers to minimize transition through the rooms. The passage of electro ducts through the walls, from one area to the next, will be centered in just one location. This location will have a perfect sealing system using injected silicone, to protect the containment area impermeability.

The lighting system of the laboratories will use fluorescent daylight lamps. They will be controlled by switches in every room. The lamp holds a reflexive aluminum lamina behind the lamps to improve efficiency.

Sanitary-hydraulics Installations

The technical decisions involving BSL-3 sewage treatment were the most discussed and which generated the most divergence among the national committee, even though all considerations took into account the norms and requirements available. It is extremely important that this system be efficient, so that potentially contaminated residues from these laboratories are treated appropriately and won't hold any risk to the population or to the environment.

The dedicated hand washing sinks were located next to the exit doors of the containment areas: one inside BSL-2 area and the other inside BSL-3. They will be controlled by foot, while the sinks in the laboratory will be manually controlled by a lever, which can be done with the forearm. These installations meet the requirements of NBR 5626. The accessories important for biosafety, such as faucet activating system and drain traps for the containment areas, also comply with the requirements of metrology institutes.

The national committee adopted the thermo-disinfection treatment for the waste water generated in the BSL-3. It is done by a boiler installed outside the containment area. This is the most recommended system, according to researchers and specialists in residue treatment, due to the low probability of microorganisms resisting such high temperatures.

Conclusion

The conclusion of this chapter presents a summary of objectives, criteria and elements used in the analysis. It also summarizes the basic characteristics of the laboratory in accordance to the analysis performed. And finally, it indicates general propositions that could be further developed from this analysis.

The main objective of this chapter is the analysis of the importance of the architectural conception and installation project, and of developing the requirement project, for the completion of the building to its finality. The other objectives include systematization and availability of information to the professionals involved in the planning of higher biosafety level laboratories. An enterprise of this kind needs procedures and application of technologies that should be understood by all involved, whether they're architects, engineers, biochemists, biologists or public administrators.

This project uses architecture and engineering as tools to minimize risks for the health of mankind, animals, and the environment. It should prevent compromises to the quality of the works developed. This is a new work process involving different agents interested in this field, such as architecture offices, public health professionals, governmental, and nongovernmental institutions.

Table 1. Synthesis of Parameters

	Component Elements of Construction	Criteria
BSL-2 and BSL-3 Containment Area and Antechambers	Walls	Resilient coating, monoliticity, impermeability and smooth finishing, resistance to chemical disinfectants, and easy cleaning
	Dividers	Rigidity, lightness, flexibility, easy cleaning, smooth finishing, easy glass or other panel installation, and possibility to install electro ducts
	Flooring and baseboards	Longevity, slip-resistance, smooth finishing and easy cleaning, impermeability, resistance to most chemical disinfectants used in procedures and cleaning, and possibility for extension to baseboards
	Ceilings	Rigidity, imperviousness, sealing when joining the wall, easy panel replacement, possibility to receive the same finishing of the wall, smooth finishing, and easy cleaning
	Layout	Flexibility, functionality Furniture resistant to chemical products and ergonomic risks, no absorbency, smooth finishing, easy cleaning, and no grooves
	Circulation/Flow	Restricted entrance control, access to BSL-2 by card and password, access to BSL-3 by password, and unidirectional flow Access to BSL-3 only through BSL-2
	Door and window frames	Smooth finishing, no grooves, imperviousness, easy handling and maintenance, resistance to chemical products, easy cleaning and disinfection Sealed windows in BSL-3 and antechambers
	Glass	Smooth finishing, low heat absorbance, transparency, easy and safe handling in case of breakage, and easy replacement
	Air conditioning and filtering systems	Inward, nonrecirculated air flow from BSL-2 to BSL-3, HEPA filters in air output, and air entry only though F3+G3 filters (HEPA filter not required)
	Electrical Installations	Outlet distribution by electro ducts, outlet with different voltages, and circuit connected to emergency supply
	Sanitary-hydraulics installations	Water handling without using the hands Sewage from BSL-3 receives treatment Sewage from BSL-2 doesn't receive treatment

While developing the BSL-3 projects, we noticed the need to adapt norms related to biosafety construction to the reality of the project. That is, the norms should be more committed to the professional reality in order to be complied with. The norms should go though reformulation as necessities and building programs change. In the case of laboratories, normatization should be improved to follow the epidemiological alterations and technological innovations that frequently occur in Brazil and in other countries. For example, when new potentially hazardous viruses appear, the need to investigate new techniques to their diagnoses and treatment may also appear. At the same time, the institutes acquire more modern equipment that speeds up these procedures. This equipment must be inserted in the physical space of the laboratory and may originate new programs that need planning of architectural solutions and installations that meet the requirements for the use of such equipment. In the same way, the industries invest in the manufacturing of novel construction materials and finishings that meet the new demands that arise from the alterations in the architectural program.

The knowledge from architects and engineers, the experience of researchers, and also specialized literature can all contribute to the adequacy of norms to the epidemiological and technological reality in Brazil. That experience, acquired in the planning of biosafety labs, can be used to show architecture schools the importance of multidisciplinarity in the development of complex architectural programs such as this. The integration of professionals from the area, like architects and engineers, is essential to turn the questions raised by researchers and technicians into architectural solutions and installations adequate to the needs of laboratory users.

Approaching these architectural solutions during graduate courses would contribute greatly to making architects become aware of operational procedures for edification adequation. Knowledge needs action, theory needs practice of researchers, architects, and engineers directly involved in complex programs. The consortium of these experiences could enrich the development of architects and engineers, not only in graduate, but also in post-graduate courses, in which it would be possible to go deeper into the problems and solutions that this multidisciplinary work can reach. This experience could also be extended to furniture designers, through knowledge of routine and procedures inside a health establishment. In the case of a laboratory, planning of furniture adequate for daily tasks of researchers would promote great benefits. Ergonomy would be better studied, and most importantly, safety in this type of building. That could result in projects of appropriate chairs specific to those activities, with measures adapted to the height of benches and procedures to be performed there. The coverings and finishings would be planned to comply with the required cleaning procedures. The information that “bench work” researchers have to pass to designers, over furniture and equipment, would help adequate design of benches, shelves, chairs, lamps and many other equipment used in the laboratories or in other health establishments.

In the near future it will be important to make post-construction assessments, which will check the occurrence of alterations during the construction relating to the project. Did the materials selected live up to expectations? Did they indeed possess the properties described by their manufacturers? Not to forget the evaluation of performance of air conditioning systems, electrical/lighting systems, and sewer treatment. Other relevant data include checking the emergency systems. Were they ever solicited? If so, how frequently and how did the devices respond? And finally, we need to retrieve information concerning laboratory maintenance, especially regarding financial expenses with maintenance and equipment replacement. The post-construction assessment would compare the performance of materials and help propose adequacies to the solutions of the project.

The completion of a Post-Occupational Assessment (POA) would hold great value for all involved in the process because this is a novel type of construction in the country. Such a report about a Brazilian BSL-3 has yet to be written. The POA would also be a valuable tool in constructing similar buildings and to improve already existing ones. The most important aspects to be evaluated at this stage include comprehensive data directly connected to the achievement of the objectives for the building of the labs. One aspect is the comparison of routine exams performed with technicians and researchers that already handled BSL-3 agents. Did the contamination levels change after the same procedures started being performed in appropriate installations?

We can't forget the fact that these BSL-3 laboratories belong to public institutes that aim at improving public health in Brazil. In this way, it would be necessary to analyze, in the future, if there were changes in the epidemiological situation in the region where they were installed. In an assignment such as this, in which numerous human, technological, and economic resources are invested, probably the greatest objective of the first Brazilian containment laboratories is the possibility of diagnosis and eradication of pathogens that afflict the population, especially those with less access to public health programs and basic sanitation.

Some laboratories, like the ones in Belém and São Paulo are practically finished. Their installations have gone through a series of tests. Some questions were not completely answered, like the difficulty of maintaining the pressure at 40 Pa. Pressure changes occur easily inside a BSL-3, especially when the doors are opened and closed continuously. That fact involves a greater complexity in air flow gauging so the pressure is stabilized near the ideal 40 Pa.

Another question yet to be answered refers to the operation of the boiler for effluent disinfection. This sewage treatment system was designed specifically for BSL-3 laboratories and there are no similar systems in other Brazilian laboratories. Regulations are still being made so that the whole system works appropriately. The water produced by the autoclave is of concern, because the amount produced has been greater than what the manufacturer had indicated.

The process for discussing the development of these projects used, as a method of analysis, the comparison of technical knowledge derived from professionals in different areas. It is only natural that, at some point, all involved acquire a basic degree of knowledge about subjects that were strange to them until then. Further studies into these multidisciplinary technical discussions will favor the emergence of new areas of occupation, different from the ones that laid the foundation for the involved professionals. The job acquires a transdisciplinary character that goes beyond the realms of domain, inviting the development of new areas of knowledge and of new professionals tuned to eclectism in their education.

References

ABNT – Associação Brasileira de Normas Técnicas.

ANVISA – Agência Nacional de Vigilância Sanitária (2000). *Normas para projetos físicos de estabelecimentos assistenciais de saúde.* Brasília: ANVISA.

ASHRAE (1993). *Handbooks of fundamentals.* New York: American Society of Healting Refrigeration and Air Conditioning Engineers.

Barker, J. H., & Houang, L. (1986). Planificación y diseño de instalaciones de laboratorio. In: Kleczkowski, B.M., & Pibouleau, R., (Eds). *Criterios de planificacion y diseño de instalaciones de atención de la salud en los paises en desarrollo.* Washington: Organización Panamericana de la Salud, Vol. 4, pp. 47-73.

Buarque, A. H. (1996). *Dicionário aurélio eletrônico.* São Paulo: Editora Nova Fronteira.

Coelho, H. (2001). *Manual de gerenciamento de resíduos sólidos de serviços de saúde.* Rio de Janeiro: Fundação Oswaldo Cruz.

Controle de contaminação. (Agusto 2002). Capela de fluxo laminar, cabina de segurança biológica e cabina de pesagem. *Revista Controle de Contaminação.* Ano 5, no. 40.

Controle de contaminação. (Novembro 2002). Tecnologias para construção e arquitetura de salas limpas. *Revista Controle de Contaminação.* Ano 6, no. 43.

Corato L. L. & Nakanishi T. M., & Caram R. M. (2001). *Inovações tecnológicas em fachadas transparentes a partir da década de 70.* São Carlos: Escola de Engenharia de São Carlos (Trabalho apresentado no VI Encontro Nacional e III Encontro Latino-Americano sobre Conforto no Ambiente Construído).

Costa, M.A.F. (1996). *Biossegurança: Segurança química básica em biotecnologia e ambientes hospitalares.* São Paulo: Editora Santos.

European Commission Directorate – General for Energy. (1994). *Daylighting in buildings.* Dublin, Irelands: Energy Research Group, School of Architecture, University College Dublin Richview Clonskeagh.

Ferreira, M. C. & Rosso, S. D. (2003). *A regulação social do trabalho.* Brasília: Paralelo 15.

FUNASA/Ministério da Saúde. (2002). Arquivos técnicos do conjunto de desenhos relativos ao projeto elaborado pela empresa Karman Arquitetura de Hospitais.

Graeff, E. A. (1979). *Edifício.* Cadernos Brasileiros de Arquitetura, vol. 7. São Paulo: Projeto Editores Associados Ltda.

Health and Welfare Canada. (1990). *Laboratory biosafety guidelines.* Ottawa, Ontario, Canada: Office of Biosafety, Laboratory Centre for Disease Control, Catalog No. MR 21-1

Houaiss, A. (2001). *Dicionário da língua Portuguesa eletrônico.* São Paulo: Editora Objetiva.

Inserm. (1991) *Les risques biologiques en laboratoire de recherche.* Paris: Institut Pasteur

IPH. (Dezembro 2002). "Reflexões e considerações sobre planejamento hospitalar e arquitetura do sistema de saúde". *Revista IPH – Instituto Brasileiro de Desenvolvimento e de Pesquisas Hospitalares.* Ano 2, no. 3.

Karman, J. (1994). *Manutenção hospitalar preditiva.* São Paulo: Editora Pini Ltda.

Lamberts, R., & Dutra, L., & Pereira, F. (1998). *Eficiência energética na arquitetura.* Universidade Federal de Santa Catarina, Florianópolis, CD-ROM.

Lima, J. D. S. (2001). *A Construção das edificações nos acidentes de trabalho: Um estudo de casos no Distrito Federal.* Brasília: Faculdade de Arquitetura e Urbanismo da Universidade de Brasília (Dissertação de Mestrado).

Ministério da Previdência e Assistência Social (1998). *Portaria n.º 1 e 2.*

Ministério da Saúde. (2002). *Biossegurança em laboratórios biomédicos e de microbiologia.* Brasília: Fundação Nacional de Saúde.

Ministério do Trabalho. (1978). *Portaria n.º 3.214, de 08 de junho de 1978.* Normas regulamentadoras que consolidam as leis do trabalho. Brasília: Ministério da Saúde.

National Institutes of Health. (1979). Laboratory *safety monograph.* Bethesda, MD: National Institutes of Health.

Oda, L. M., & Ávila, S. M. de, (orgs.) Ministério da Saúde. (1998). *Biossegurança em Laboratórios de Saúde Pública.* Brasília: Fiocruz.

Palmer, C. (1976). *Ergonomia.* Tradução de Almir da Silva Mendonça. Rio de Janeiro: Editora Fundação Getúlio Vargas.

Panero, J., & Zelnik, M. (1984). *Las dimensiones humanas en los espacios interiores.* México, D.F.: Editora Gustavo Gili.

Richmond, J. Y. (1999). *Anthology of biosafety: perspectives on laboratory design.* Mundelein, IL: American Biological Safety Association

Richmond, J. Y., (1997) *Designing a modern microbiological/biomedical laboratory: Lab design process & technology.* Washington: American Public Health Association.

Superinteressante. (Fevereiro 2004). "Presídio viral". *Revista superinteressante* Edição 197.

Teixeira, P., & Valle, S. (1996). *Biossegurança: Uma abordagem multidisciplinar.* Rio de Janeiro: Editora Fiocruz.

Valle, S. (1998). *Regulamentação da Biossegurança em Biotecnologia.* Rio de Janeiro: Gráfica Auriverde.

Veja. (Maio 2003). “Uma epidemia globalizada”. *Revista Veja.* Ano 36, no. 18.

World Health Organization. (1993). *Laboratory biosafety manual,* 2ª Ed. Genebra: Author.

Chapter 4

Design Considerations for Large Scale Production of Biologicals: GMP and Containment Synergies

Vibeke Halkjær-Knudsen, PhD

Abstract

The biosafety level for production of polio vaccine is being upgraded as Poliomyelitis is close to being an eradicated disease. This necessitates upgrading IPV (Inactivated Polio Vaccine) facilities around the world to a higher and more stringent biosafety level. Merging Good Manufacturing Practices (GMP) and high containment regulations is necessary, and a decision must be made whether to build a new facility or upgrade the existing one. Biosafety guidelines and GMP rules basically support and reinforce each other, however, in some cases they contradict. When working with high-risk agents, totally new vaccines, or soon to be eradicated agents, containment must be achieved, but product safety must not be compromised. The main tool to weigh out contradicting recommendations seen from a product-GMP, biocontainment, environmental, and personal safety point of view are risk assessments.

Keywords: GMP, biosafety, risk analysis, facility design, large-scale production, equipment design, design of utility systems, kill system, decontamination, fumigation, IPV, Poliomyelitis, Polio, high containment.

Introduction

Aseptic production of pharmaceuticals usually takes place in a positive pressure environment to keep particles and microorganisms out. Production of biologicals under Bio Safety Level 2 or 3 (BSL-2, BSL-3), or higher requires that production facilities are operated with a relatively negative pressure in order to keep viruses and bacteria inside the facility. The biosafety requirements for a vaccine GMP production is fairly low (BSL-2), the level observed until now. Recently, however, the biosafety level for production of polio vaccine is being reconsidered. WHO expects and plans a worldwide eradication of Poliomyelitis in the near future, which necessitates a higher containment level for all polio vaccine productions around the world. We have analyzed problems and synergies of merging GMP and containment aspects in order to find the best possible way to construct our new facility and still comply with current guidelines. In the following article, I will share some of the considerations made at Statens Serum Institute, Denmark, during the planning and construction of our new large-scale polio vaccine facility.

Planning a GMP production usually involves large rooms, large air volumes, and at least 20 air changes per hour in the clean rooms. Biocontainment work is usually carried out in small laboratories with no clean room specifications or demands. At first glance, trying to merge GMP and containment guidelines may seem to be an almost insurmountable task, however, when working with the two 'worlds', they seem to converge.

Definition of Biosafety

Biosafety is a combination of procedures, containment systems, and construction technologies designed with the purpose of reducing the risk of infecting laboratories and releasing microbes into the surrounding environment.

Purpose: The purpose of biosafety is to create a safe environment when working with contagious diseases, i.e. to prevent release of contagious agents, to minimize employees' and other people's contact with the agent, both inside and outside the containment zone, and to prevent the introduction of contagious agents into the environment.

Definition of GMP

The Good Manufacturing Practices (GMP) is the part of the quality assurance system that ensures that pharmaceutical products are produced consistently and controlled in accordance with the appropriate standards for quality. These depend on the intended use of the product and the requirements issued by the marketing authorization or the product specification. GMP applies to both production and quality control.

Purpose: To ensure that the product is safe for the end user.

Biosafety Principles and Characteristics

It is relevant to think in terms of biosafety when producing or analyzing vaccines, when using GMO's (Genetically Modified Organisms) in production, or when working in research laboratories with biological risk material or GMOs. In regard to hospitals, it is relevant when designing isolation suites for patients with exotic diseases. Biosafety is also relevant for some animal facilities and when handling items associated with bioterrorism.

The categorization of bio hazardous agents is almost the same around the world. Exceptions are some diseases classified as exotic in one part of the World, but found naturally in the environment in other parts. These diseases might be considered less dangerous in their native habitat and so given a lower classification there, but they will still be given a high classification in the rest of the World.

The biosafety level for a specific agent is defined using the relevant national or international guidelines; find the agent in the tables, and perform a risk analysis. Biosafety is a combination of construction methods, process controls, administrative controls, working procedures, PPE (personal protective equipment), and evaluations of health hazards. Due to scale, handling, and the possibility of an incident/accident risk analysis might give reason to handle the agent at a higher containment level than suggested by the guidelines. Specific decisions regarding construction, design, containment details, and choice of equipment can be made based on this.

Some of the biocontainment guidelines are focused on the end result and leave it up to the individual to decide how the goal is best achieved, including level of commissioning, qualification or validation. Other guidelines are more focused on the process and specify in detail how the goal should be achieved, including the acceptance criteria for this.

There are 4 levels of biosafety:

BSL-1

Defined organisms and contagious agents known not to cause disease in healthy adults. There are no specific requirements regarding containment or otherwise for these agents.

BSL-2

Moderate risk agents found naturally in the environment and the population. The agents are known to cause disease, but the severity of the progression of the disease may vary. A cure or prevention by vaccination might be available. There are no specific requirements regarding containment or otherwise for these agents. It is, however, relevant to consider procedures and vaccination status.

BSL-3

Exotic agents known to cause serious and life-threatening infections in humans. Aerosol spreading is a risk. The laboratory or production area should be constructed as an isolated zone with an anteroom or airlock and easy access to hand washing facilities. Considerations regarding exhaust air are very important. Another relevant issue is the safe handling of infectious waste. Decontamination of the waste must therefore be addressed. This is extremely important, especially for large-scale productions, where due to the scale, it is almost impossible to decontaminate everything in an autoclave.

BSL-4

Dangerous or exotic high-risk agents known to cause life-threatening infections. The laboratory or production area must be in a separate building to obtain the required extreme level of isolation. All waste must be handled correctly and space dedicated to support/back-up systems is needed. Decontamination facilities are very important and must be well designed. Double door autoclave, kill system, HEPA filtered supply air, double HEPA filters on exhaust air, showers, space suits or isolators are some of the aspects that must be addressed. The items listed above are just examples. To get a full view of all aspects, the relevant specific guidelines must be consulted.

Levels of Containment

In biocontainment terminology, there are three levels of containment:

Primary Containment

Addresses containment of aerosols and splashes. It is the barrier between the agent and the technician. It can be the vessel, the HEPA filter or biosafety cabinets, isolators, spacesuits, etc. For a large-scale production, the primary containment barrier is normally the fermentor.

Secondary Containment

The room, the laboratory, and the various technical installations. It provides a barrier that protects people outside the laboratory from being exposed should an accident happen inside the laboratory.

Tertiary Containment

The building and procedures. It adds to the barrier level towards the community for unintended exposure during accidents.

Containment barriers are achieved by physical separation, practice and procedures, decontamination, and filtration.

Containment barriers are broken by people, air, gasses, liquid waste, materials, and products.

Having a well-trained and educated staff is one of the most important factors in achieving effective containment. They must be continuously trained to remain alert against risks and comply with safety procedures.

Risk Assessment

Large-Scale Production – Is It Risky? Typically a production area is larger than a research or diagnostic area, and the volumes of hazardous materials are also much larger. A large-scale production adds risk factors that must be considered as well as those seen for a normal laboratory. Risk analyses should be carried out for steam, hot WFI (Water for Injection), compressed air, gasses and chemicals for CIP (Clean in Place) of storage tanks, and fermentors. These analyses further highlight the aspect of emergency exits. The rooms contain a lot of utilities, all of which pose a risk to the employees, and the OSHA, therefore, recommends emergency exits in case an accident should occur. However, for containment reasons, as few doors and exits as possible are warranted. A risk assessment creates balance between all these viewpoints.

Large amounts of highly infectious material is handled, and pipes carrying large volumes of hot WFI, gasses, waste, pure steam, CIP chemicals, contaminated waste stream, acid, and base can be found throughout the production facility. Should an accident happen in a large- scale facility, the consequences will be much bigger than if an Erlenmeyer flask is dropped on the floor in a small-scale laboratory. On the other hand, closed systems are used due to GMP process demands — double air filters on vessels and steam traps, etc. to keep microbial contamination out and maintain the pathogen inside the vessel.

The closed system also includes sterile connecting devices for inoculation and sampling, surveillance systems, alerts, alarms, and automatic shut downs in response to critical alarms. And last, but not least, the staff follows GMP procedures, which means a GMP-trained staff, batch records, SOPs (standard operation procedures), logbooks, validated procedures, etc.

The risk assessment should address and document these aspects as a minimum. The assessment must take the following items into consideration:

- **The reservoir:** Is it a stainless steel tank, a test tube in a centrifuge or an Erlenmeyer flask? How can it break? What is the possibility that it will break or leak?

- **The volume:** Is the possible spill a few millilitres or more than 1,000 litres?

- **The concentration:** What is the concentration of the sample?

- **Way of escape:** How can the spill get out? The centrifuge may crush the vial, the bottom valve on a fermentor might leak, or the filters on the fermentor might malfunction.

- **Way of transmission:** Is it airborne or not? Can it fly by itself or does the agent need an aerosol to be generated in order to move around?

- **The route of transmission**: Does it infect directly through the lungs or the skin? Or does it have to be introduced into the blood stream before it becomes dangerous?

- **The infectious dose**: How many infectious units are needed to get sick? Is one enough or are a couple of thousand units needed before someone will get ill?

- **A susceptible host:** Who will get sick? Everybody? Is a vaccine available? Are immune compromised people at specific risk?

- **The incubation period:** How long is the incubation period? Is treatment effective if started during the incubation period?

- **Decontamination principles:** What chemistry will kill the pathogen? What kind of "problems" is associated with the chosen chemical for decontamination? Is it carcinogenic? Is it poisonous? Does it give rise to allergic reactions for the employees? Is it a problem for the environment?

Can a spill be handled with chemicals directly on the floor? Maybe if the spill is relatively small, but what if the possible spill is 2,000-3,000 liters? How many liters of decontamination liquid will this necessitate? How can a full and effective decontamination be ensured in this scenario? Might the spill be increased - maybe because of a broken pipe with running water, diluting the agent but enlarging the volume?

These are examples of some of the questions that should be addressed during a risk assessment for every agent, laboratory or production facility. The answers and the design consequences will differ for every site and every user. Some can build a lot of safety into the building and the equipment itself. Others will need a lot of procedures or protective equipment to accomplish an acceptable level of safety.

This is why there is no easy way of setting up a universal set of rules in biosafety that will ensure that if these are just followed, or these materials used, or these building principles implemented - then you will be safe.

You need to do your own risk assessment and decide which solutions, materials, and designs will suit the purpose best. This should be done very early in the project. As long as the project is still on paper, materials and surfaces can be changed for a reasonable sum of money. It becomes proportionally more expensive as the project goes on and the facility is being built.

GMP and Containment

Where do the guidelines reinforce each other and where does their focus diverge?

Synergies

Some of the areas where the GMP and containment guidelines reinforce each other and synergy is experienced, are listed below:

- Easy, cleanable design and a focus on minimizing contaminants.
- Validate all processes, systems, equipment, and utilities.
- Isolate production areas from other areas, and establish some form of admittance control.

- Provide mandatory, certified, and documented training as well as mandatory personal. protection equipment/PPE.
- Revise written policies and procedures at fixed periods.

Conflicting Areas

The conflicting areas are primarily room pressure and material flow with the focus on particle and microbial environment control: Keep out (GMP) or keep in (BSL)?

- GMP guidelines specify that an HVAC design with a positive pressure is preferred in order to avoid cross contamination. However, the containment guidelines view this from a different angle and specify a negative pressure due to environmental issues.

- Material flows from clean to unclean are also areas where conflicts arise. The definition of clean and dirty differs for the two areas:

  GMP: Raw materials (dirty) to end product (clean).

  Containment: No contagious (clean) to contagious (dirty).

This complicates the handling of materials, personnel flow and waste flow, and may increase the amount of cleaning carts needed in the facility as the rooms should be cleaned from "clean to dirty" in both worlds.

Which guideline "wins?"

Health risks versus product risks must be evaluated before a decision can be made, but in general, the rule of thumb is that working under BSL-1 or BSL-2*, the GMP rules have the highest priority due to low health hazards. However, at higher levels of health hazards, BSL-2*, BSL-3 or BSL-4, the bio safety guidelines have the highest priority. *Depends on the actual agent, scale, etc.

In reality, both types of requirements should be met when working with a large-scale BSL-2/BSL-3/BSL-4 GMP-production. GMP and containment authorities will inspect the facility, and both will expect to see design choices according to their standards. Most of the time synergy will exist between the two sets of guidelines; however, some contradicting issues may arise trying to merge the two worlds. It is very important to document in the DQ (design qualification) and the risk evaluation why one set of guidelines is chosen above the other, or why a specific aspect is being addressed using different methods than normally seen. Note that OSHA requirements and personnel safety must also must be addressed.

Design for GMP and Biocontainment

Design considerations must be addressed at the earliest time possible, and decisions must be made in those areas where GMP and containment do not reinforce each other. It is wise to document the rationale behind specific design decisions in a format and in a place where it will be possible to locate it at a later time when suddenly needed during an inspection. Below is a list of some points to consider.

Building

Whether it is a new building being built from scratch or a reconstruction of an existing building, whether it is for containment, GMP or both, several questions must be answered before a new project is set in motion.

Does the building have to be completely separated from other buildings or is it acceptable to place it close to existing buildings, if it is ensured that all entries and access points are completely separated from the surrounding area? Is it acceptable to share existing utilities with other buildings on site, etc.?

What level of containment and security is needed in- and outside the building? Bullet- or bazooka-proof windows, open doors during daytime, card access, iris scan, finger prints, camera surveillance (CCTV), or something else? If choosing card access, will ID badges need to be autoclaved before leaving the containment zone, and can the cards withstand that kind of treatment over time? If choosing finger print access - what is the dress code? Should employees wear gloves when in the production rooms?

Do employees have to work in shifts during evening and night hours? Is there a guard on site? Should alarms go to an external security company? To what level should the night watch be trained to address emergency situations? The operation of a large-scale production facility during night hours is very different from the operation of a diagnostic BSL lab. Choosing the wrong strategy for how to handle an emergency stop or breakdown during a crisis, the wrong method to try to minimize the effects of an accident, or the wrong way to start the facility back up, might all have disastrous consequences from a containment point of view.

What kinds of emergency exits are acceptable? Is it important from a containment point of view that emergency exits can be closed behind employees leaving the building during an emergency? Are "hammer out" windows an option or are doors with a rip-off sealant preferred thereby assisting in keeping the contagious agents inside?

Is an explosion within the facility a risk to be addressed?

What is the size of the equipment that might need to be installed in the distant future, and how big should the openings in the construction be to allow for this?

Which materials are preferred for the construction of the building? Will it be necessary to bring new equipment into the production rooms on short notice? Walls made of glass and stainless steel generate much less dust and fewer particles (and can probably be reused) when dismantled, than walls made of concrete or gypsum.

The size of the new production facility might increase steadily during the initial design and programming phases, as more knowledge is gained. 3-D modelling is very helpful when designing for serviceability and placement of manometers and manual valves. Looking at the equipment from above often gives a completely different picture than originally imagined. Another helpful supplement to 3-D modelling is "the stone age method," i.e., building part of the production area and equipment on a 1:1 scale using cheap materials. One way to do this is to remodel existing equipment (CIP units, mobile tanks, etc.) by attaching the right type of wheels, adding the right load and modifying the equipment to fit the dimensions of the planned new equipment. Place the 1:1 equipment in a large empty storage hall, paint stripes on the floor indicating the location of walls etc., and let personnel test the dummy equipment on this scale to see if there is enough space to move it around. Have personnel wear PPE at some point during the drill, especially if they might have to wear it for real emergencies at some point - for example to protect themselves while cleaning up after a major spill. Moving around with a couple of bottled air tanks on the back might change the opinion of how much space is needed. Recognizing this issue during the design phase is a lot cheaper than moving walls later on in the project.

Is it necessary to fumigate rooms routinely or only after major spills? Room tightness and surfaces become specific points of interest. Should the technical areas supplied with pilot valves and electrical cabinets be open to allow easy fumigation or closed with an IP (internal protection) number high enough for you to feel confident that no contamination can stay inside and wait for an opportunity to cross-contaminate at a later time?

Flows

The flows of materials and employees must be designed to ensure that mix-ups, mistakes, and cross-contamination are avoided. During the design phase, this aspect requires extra consideration, as the two definitions of "clean and dirty" must be addressed. A normal GMP flow never moves backwards because "dirty staff" is not supposed to meet "clean staff". In the containment world, however, labs are usually constructed as a "blind end" where personnel enter and exit through the same door.

Surfaces

Surfaces, equipment, and utilities should all be designed in an easy cleanable way, i.e. smooth surfaces and seamless floors and ceilings. Equipment with nonporous surfaces able to resist cleaning agents, disinfection, and decontamination are preferred whether building for GMP, containment, or both - the demands are the same. How are walls and doors protected? Fender lists: dimensions, height, strength, material - are these materials resistant against corrosion created by cleaning agents and disinfectants?

Should the surfaces be painted with a diffusion tight paint, coated with epoxy or similar materials? Should the walls be constructed as stainless steel or vinyl covered concrete/gypsum? Are glass or Arcoplast walls a possibility? Should tightness be achieved by gluing, welding or sealing?

What is the curing time when using concrete and how long do you have to wait before construction can continue? What is the drying time for some of the other materials? How tight is the building schedule?

Can glass walls assist in heightening the safety around the staff? Glass can be made break resistant to a degree, which makes it comparable with other types of materials. Compared to the view into an adjacent room through a small window, glass walls make it very easy to gain a full view of what is going on in rooms throughout the building. Should a colleague get sick or an automatic process run loose, it will be observed immediately. When cleaning the glass walls with steam and wiping off afterwards, cracks in sealants will be discovered immediately, as the steam will go through the wall and condense on the backside of the glass.

Is color coding preferred on floors for clean rooms, labs and containment zones? Do the chosen colors withstand normal cleaning/disinfectant agents? Test it. Do not exhibit blind faith in anyone who presents a product specification on a sheet of paper. Robustness of pigmentation is as relevant as the actual material resistance.

Is the floor material strong enough to withstand the resulting wear and tear from rolling equipment? Test it using wheels of the right dimensions and the right material.

Will it be necessary to repair surfaces while the production is still running? Will maintenance periods be scheduled where these aspects can be addressed, or is it preferable to do preventive maintenance while production is going on? These answers may also be relevant for the choice of materials.

All materials have both pros and cons in durability, acoustic profile, maintenance, repairing time, and price. What is found to be the best choice in one scenario might not be the same in another.

Interior Design

Both containment and GMP guidelines state that lamps must be installed in a way that makes them easy to clean. What level of lux/candela is needed? What about animals or cell cultures? Do they need reduced lighting? Would it be desirable to be able to increase the lighting during maintenance compared to a normal production situation? Should touch-free on/off switches be chosen? What technology does this involve? Heat sensitive (37°C) switches will, surprisingly enough, not function in incubator rooms! Are unbreakable fluorescent tubes and laminated glass fronts on the lamps in the ceiling desirable from a safety point of view? Should the lamps be serviceable from above?

Emergency lights should be installed on the ventilation floor after all equipment has been installed. This strategy reduces the number of lamps to be remounted later on! Can normal lamps be used? Or would special emergency lamps be better? Should it be possible to bring some of them with you, like a pocket lamp?

A switch close to the entrance, to turn all lighting in the entire building on/off at the same time, is a nice feature. It will be appreciated when the staff leaves the light on in the production area. It takes a lot of regowning and showering just to turn off the light. When getting an alarm from the facility in the middle of the night, a central on/off switch is also an appreciated detail.

Should heat or ionic sensors be the choice for automatic fire sensors? Ionic sensors also react when hot steam escapes from washing machines, autoclaves, leaking valves, or during SIP (Steam in Place). How close to the production equipment should the sensors be placed? How far away is it wise to place them? Is it possible to place them in a location where they can be inspected from no containment areas in adjacent rooms? Where should windows be placed in walls and doors to accomplish this? What window size is necessary? Do fire fighters accept lying on the floor of a room right outside the containment zone to get the right viewing angle in order to inspect the fire sensor in the next room?

Does building code prescribe automatic sprinklers to be installed? Both from a GMP and containment point of view, automatic sprinklers should be avoided. From a GMP point of view, because they might contain unclean water and thereby create a problem in clean rooms and from a containment point of view because the water might multiply the volume and move the spill from where it initially was, to a much larger area.

Are there locally-trained fire fighters on site? Invite the internal and external fire fighters to a meeting during the design phase and discuss these aspects thoroughly. What kind of suit should they wear going into the facility? Should they strip off their clothes for decontamination on site, before the clothes are brought outside? Will this procedure fit in to existing procedures for the fire fighters? Can the suits withstand decontamination? These aspects were quite new to our city fire fighters, as was the discussion about how to extinguish a fire safely in a containment facility. We prefer that they do not use large amounts of water and that fire hoses are not used inside at all, as these will keep doors open and break containment.

Must the floors of the rooms serve as a basin during spills? Do spill handling procedures prescribe adding a decontaminant to the initial spilled volume for *in situ* inactivation of the spill? What is the volume of the largest tank? Is this fermentor or tank linked to other tanks or pipes at any time during the process, and could the spill be diluted, thereby creating a larger volume than originally expected? How is the spill removed afterwards? These procedures should be tested as early as possible, and the commissioning timeline should allow for a redesign phase if necessary. Is there large equipment on wheels that needs transport across containment perimeters, and what is the acceptable slope for the needed ramp?

Technical Space

Paint stripes on the floor indicating the walls of the rooms below before installing equipment. This makes it easy to locate the right lamp when a light bulb needs changing. Put fluorescent stripes on the floor indicating the quickest route to the emergency exit, after installing all the equipment and in agreement with directions from the fire fighters. Place the electrical conduits above the pipes leading liquids throughout the building - logical, but there is no harm in pointing it out to the contractors. Waterproof floors are a very nice detail when having a clean room production beneath. The extra investment might be justified sooner than expected! Storage space for extra filters is worth considering - they do take up a lot of space. Separate the containment ventilation from the rest of the aggregates, if possible, and consider whether this part of the technical space should be possible to fumigate as well.

Room Tightness

How tight should the rooms be? Consider aspects such as containment and fumigation. Is fumigation expected to be initiated after a specific incidence, e.g. a larger spill, or routinely?

If absolute negative pressure compared to a relatively negative pressure is chosen for the production area, the rooms need to be tighter, as inflow of dust and dirt into the clean rooms through the construction crevices becomes an important issue.

Using contrasting colors for sealants compared to building parts will help to detect unfilled holes and crevices during visual inspection. Pressure testing the containment zone is a good method, as the result can be used for comparison during recommissioning at a later time. BSL-3 facilities do not need to be constructed according to BSL-4 requirements regarding tightness and do not need to comply with, for instance, the Canadian acceptance criteria of 12, 5 Pa/min pressure loss. It does not change the fact though, that it is a method that will show whether the building characteristics have changed over time, after a reconstruction phase or whether the perimeter integrity is undamaged. However, less strict acceptance criteria should be chosen. Another way of using pressure differential to test room tightness is creating a high pressure differential between the two sides of a building part, releasing dirty air or particles on one side, and measuring with a particle counter on the other. Using a suction cup with vacuum and a gauge on a smaller section of a construction part is also a fast and easy way of testing the actual tightness of sealants. Methods such as smoke, smelling substances, helium, and soap bubbles are also used. The detection limit varies for the different methods.

Consider dividing the containment area into smaller separate sections which enables part of the containment zone to be decontaminated or tested for tightness, while the rest of the production is still running. How is negative pressure created? Should some of the doors or parts of the ceilings have plugged holes with specified connectors where vacuum generators can be mounted?

How are the electrical switches tightened? A gastight design should be considered from the beginning. Polyethylene is a nice, PVC-free material, which helps to preserve the environment but it is almost impossible to find a sealant that will create a tight enough seal between the two surfaces to comply with the specified acceptance criteria.

Doors and Airlocks

What kind of doors are needed, airtight, normal doors or doors with a sweeper along the floor? Airtight doors are relevant if routine fumigation is planned. However, if this is not the case, taping or sealing the doors when needed could be considered instead, as airtight doors are heavy and maintenance demanding (whether pneumatic or with gaskets). After all, they still have to be opened to get in and out of the rooms. A sweeper beneath the door will add to the tightness, but might be maintenance demanding for optimal functionality.

Is interlock needed for the airlocks? This is preferable for a GMP production. What is the cleanup time? From a GMP point of view, it is obviously an issue to make sure that dirt from the corridor stays in the airlock and is not brought further into the clean rooms. From a containment point of view, however, the cleanup time refers to the time from when one person leaves the airlock and the next one enters. Here, issues such as turbulent air from the containment clean room production flowing backwards into the airlock should be addressed, ensuring that the airlock is as clean as possible and that nothing can flow further backwards out of the containment zone. All these issues should be addressed and tested. Particles can be generated to simulate air from a heavily-contaminated production room flowing backwards into the airlock to get the right picture. This is why ingoing airlocks always need to be served by the containment ventilation system that is supplied with HEPA filters on the exhaust.

From a containment point of view, hand washing facilities should be placed in the production rooms. However, running tap water might not be the best choice from a GMP point of view. A compromise could be to put them in the airlocks - a space-demanding detail to be remembered during the design phase.

Are the airlocks to and from the production rooms large enough to hold the emergency personal protective equipment that might be needed during a large spill on a permanent basis? Normal containment procedures prescribe that personnel must leave the room and let the aerosol settle for at least half an hour before going back in to handle the spill. Working with a large-scale production is different. Here the priority is to contain the accident and stop it from spreading further. Heating mantles, steam valves, pumps, and most large-scale equipment can increase the damage, if not stopped immediately. Personal protective equipment is therefore needed close by, preferably not in the actual room where an accident can be foreseen to happen. Storing it in the airlock is a possibility.

At what height should door handles be placed? Is elbow contacts or draglines from the ceiling preferred? Elbow contacts are very desirable from a containment point of view.

Which way should the doors open? Against the largest pressure, as the elevated pressure helps to keep the door closed? Should it open into the largest room, as it lessens the turbulence and the backflow of contaminated air into the smaller airlock? Outward movement of potentially virus-contaminated air might cause a containment breach. If the door closes against a larger pressure in a small room, will a tight-fitting door jump back up, because there is not enough space in the room for all the air? How fast are the air changes in the airlock? Is the airlock equipped with both ventilation inlet and exhaust? How fast will the door be closing? A slow opening and closing door will give fewer problems, but might frustrate personnel on a daily basis. In corridors with a lot of traffic, the doors should always open away from traffic. During normal working hours, personnel should not have to move towards a door, which might swing open and smash into their face and the things being carried, or the rolling equipment being transported along corridors. Does the door open into a very small room filled with cleaning carts or other stored items and will there be enough floor space left for these carts? People entering from the corridor to retrieve the cleaning carts and closing the door afterwards will mainly use this door and will therefore not pose a risk for hitting a colleague. This door might add value to the small room by not opening into the crammed space but outwards into the corridor. Finally, there is the fire egress issue, which prescribes that it must be easy for people to exit the building in case of an emergency. Which way should the doors open in an emergency? Sometimes these demands will coincide and sometimes they will not. A specific decision must be made for each door regarding size, tightness, and the way it opens. Make sure the architects have all the necessary information about the specific demands of each door. Otherwise, they will be unable to make the right decision on their own regarding which way to swing the door. They might address it from an aesthetic point of view, all doors swinging either in or out of the corridor, as this looks very nice on a drawing.

The airlocks will be supplied with step-over-benches due to the clean room GMP production aspect, which means that personnel must jump or step over the benches while leaving the building; this is definitely not optimal from an emergency point of view. Emergency lights must therefore be placed to ensure that step-over-benches and other obstacles are illuminated adequately. This solution can be considered acceptable, mainly because when having a GMP-production a documented small and highly skilled, well-trained staff will be ensured. They will be used to working in this kind of environment, where they might have to leave in a hurry at some point and they will know their way around.

The GMP guidelines specify that airlocks must be dedicated specifically to either transportation of materials or personnel; airlocks cannot be used for both. From a containment point of view, however, it is preferable to have as few "holes" as possible into the area. Again a risk analysis will be the deciding tool, as this will show how much an airlock dedicated to materials alone might compromise the containment aspect. In a large-scale production, a lot of the materials needed are piped through the walls, instead of transported around in mobile tanks. Is building a dedicated material airlock worth it, if only about 500 ml of virus inoculate is brought in twice a month, considering the amount of work it requires to seal off the perimeter, if fumigation is needed? If you build a material airlock, should it be constructed as a dunk tank with a disinfectant? What is the possibility of the surface being contaminated with the items being brought out from the containment zone? How large should the dunk tank be, according to the best guess from the staff? Double the size — even if it takes more disinfectant! Should some of the material airlocks be designed as full-scale fumigation chambers, as a supplement to autoclaving contaminated material out of the facility?

Chemicals for Decontamination

Which ones will suit the purpose? Is it enough to wash down with a disinfectant or to clean with steam, or is fumigation with formaldehyde or peroxide the only acceptable option? It is a tremendous task to wash down large production equipment using normal washing procedures due to the complexity of the equipment, electric wiring, etc. Every procedure and chemical must be considered when choosing construction and surface materials, as they might influence the decision; some chemicals bleach the surfaces, some materials get sticky when exposed to harsh chemicals and some types of sealants become foamy and start to bubble when exposed to a specific sequence of disinfectants over time.

Fumigation

Which agent is preferred? Formaldehyde, peroxide, or another chemical? Should concentration be monitored during fumigation? One way to do this is to design the pressure monitors mounted in the ceiling as T-valves. They can then be hooked up to the analytical instrument during fumigation, and the specific room concentration can be monitored in a safe way, without having anybody enter the room. What are the acceptable levels of exhaust from the environmental authorities? Should the fumigation include the ventilation ducts? Should it be possible to use the ventilation system to help homogenizing the concentration throughout the area? Is fumigation of the whole area or just individual rooms planned? Should it be possible to continue working in adjacent rooms? Can the fumigation equipment be turned on and off from the outside? Are the switches on emergency power? Where are the chemicals for an emergency fumigation stored? Will the fumigating agent corrode any of the surfaces? Will the hexamine complex that might be formed by the formaldehyde neutralizing amino compounds, harm the electronic equipment? Is a fixed concentration of the agent needed during fumigation? Will the compound degrade and does it have to be replenished? All these questions must be considered before closing the design discussions in the programming phase.

HVAC

The construction of the building, the utilities, and the HVAC design should focus on minimizing the possibility of contamination and cross-contamination.

Should HEPA filters be changed from above (technical space) or below (production room)? How tight should the ducts leading from the room to the HEPA filters be?

Should automatic or manual dampers be chosen? This might depend on the safety systems (emergency shutdown) and the planned fumigation schedule. In and out-going air should be coupled to ensure the specified pressure differentials.

Where is the ventilation exhaust placed? How is the ventilation shafts secured in order to prevent intruders from gaining access to the building through these? How about one-way screws?

Is the technical space vented passively or is a specific system needed? Will extra ventilation be needed during formaldehyde fumigation to flush the areas if something goes wrong? Or is the design so spacious that people can wear PPE and work freely while fumigating?

When all tests have been performed and commissioning is finished, it might be wise to check the ventilation ducts for newly-drilled holes made by the latest testing crew, who just needed a supplemental airflow measurement. Once refilled, these holes must be checked frequently during the yearly maintenance. They should therefore be marked with numbers fully visible on the duct and on a drawing.

In a clean room production, air changes above 20 air changes per hour are preferred, and in order to obtain a constant air change due to GMP specifications, the design can be set up with a constant inflow of air and a variable exhaust flow to regulate the pressure. In this case HEPA filters should be positioned in front of the regulating devices, as these devices otherwise would have to be decontaminated before preventative maintenance.

Where should exhaust vents be placed? In the ceiling? Or should vertical shafts running down close to the floor be considered? Will it be necessary at some point to supply these vertical shafts with HEPA filters on the end within the room in order to protect the rest of the exhaust duct from being contaminated?

What kind of pressure differential is recommended in guidelines across the containment perimeter? The larger the differential, the more difficult it will be to open and close doors.

Test and document the ventilation system as if it was a normal GMP system:

- **IQ (installation qualification):** 100% component documentation and labelling to ensure replacement of an identical part. SOPs (standard operating procedures) for production, maintenance, start-up and shut-down procedures, changing filters, decontamination before maintenance, etc. should be written. Drafts of these SOPs would be helpful if attached to the user requirement specifications as appendices. Knowing these procedures from the start would make the design and construction phase easier.

- **OQ (operation qualification):** Documentation of air changes, pressure, HEPA leakage rate, clean room classification, humidity, video of flow patterns, and pressure gradients from clean to unclean. As the definitions of clean and unclean are different in GMP and containment terminology, both flows should be addressed, verified, and documented. It is also relevant to test if the chosen diffuser design is acceptable for personnel working on the various raised platforms through-out the building. If the initial design creates too strong a draft for personnel working on the platforms, it needs to be changed. This test should be done early in the documentation and verification phase, to prevent other commissioning test, to be redone after diffuser change.

When testing for an unplanned shut-down, it could be a good strategy to list the ventilation systems in a matrix. Test by taking one exhaust or inflow system offline at a time, and verify that the rest of the systems behave and interact as planned and specified. Start the test by placing all doors in a secure open position. When the test has been carried out successfully with open doors, the test can continue with the doors being closed successively. Whether this strategy is possible depends on the design of the system, and how alerts and alarms will interact. The reason for starting the test with open doors is obvious. Should something not act as specified and the ventilation system behave in an unexpected manner, it is possible to handle the situation safely and not tear the building apart. Remember to open the doors to the outside of the building when starting up the test and do the test in stages. The test might take much longer, but it will be worthwhile. If the pressure goes beyond specifications, the ceiling might end up on the floor during the test, and a lot of time is saved not having to reconstruct the ceiling. Many walkie-talkies are needed to keep all test personnel, whether situated in the basement, on ground level, top space, and in the data center updated and to coordinate the efforts during the test. Walkie-talkies using the same transmission band are needed.

Some constructions have a built-in mechanical pressure relief, which allows air to escape through the walls into the neighbouring rooms through a u-tube containing a disinfectant or HEPA filter. The tube containing disinfectant must be replenished from time to time with an acceptable chemical - acceptable both from a disinfectant and personnel safety point of view - as the employees might get hit by the spill. Another solution is to build in a simple construction in the ducts behind the HEPA filters, which opens and closes letting false air get sucked in using a weight as a counter balance. It is a rather inexpensive solution, but precious time will be lost if it has to be added to the design at one of the later stages of the project. Experience has showed that automatic process control systems might not be enough in an extreme situation.

HEPA filters might need to be bagged out or decontaminated before being replaced. It is important to make sure there is enough space around the ductwork for a formaldehyde fumigation unit. Adequate workspace in front of the filters is also necessary. Both these issues must be addressed early in the programming phase. A strategy for DOP (di-octyl-phthalate) testing filters that might be contagious must also be addressed.

Equipment and Utilities

Equipment should always be placed with an adequate amount of space around it or be sealed to the floor or ceiling in order to make routine cleaning or gassing possible. For very large equipment, for example autoclaves, it is a good idea to build in an extra safety barrier. If for example, the epoxy coated concrete beneath the autoclave shows a tendency to crack, how will it be repaired? It might be more convenient to install the autoclave in a welded stainless steel bathtub from the beginning.

Wheels are a necessity under some types of equipment, e.g. media containers; however, the available selection of easy rolling and autoclaveable wheels that will not leave black marks on the floors is very limited. Only specific dimensions are available.

Sampling valves must be placed in an acceptable position, both from a microbiological and ergonomically point of view. As few connections as possible between the various parts of the production equipment is preferred, however, this also results in reduced equipment flexibility.

Equipment noise should be reduced to a minimum using noise shields, however, these shields must be either aerosol tight or easy to demount for fumigation. This is also relevant for other demountable insulation parts on, for example, pipes, e.g., with hot or cold media.

Heat resistant TAG numbers are very important. In an emergency, precious time might be lost looking for the correct valve to close if a TAG number is missing or has become unreadable.

Pipes that are SIP'ed (steamed in place) at intervals can be marked with heat sensitive labels that change colour, as the heat increases. This increases personnel safety, as people tend to forget to be careful around pipes that are only hot part of the time, thereby causing burns on arms and the backs of hands.

The heat expansion aspect needs to be addressed for piping going through the containment perimeter, if these pipes are SIP´ed from time to time. The penetration through the wall must to be tightly sealed whether the pipes are hot or cold.

Gas pipes should be mounted with a slight slope towards the containment zone. This will make it possible to drain them from inside the containment zone after they've been decontaminated with steam. This is a must should a backflow occur from the fermentor into the pipes.

Drilling holes in walls or ceilings should always be done from inside the containment zone. This will make them easier to seal.

The ability to disconnect compressed air to specific pilot valves is relevant, if containment aspects and a risk assessment points out that a software malfunction or erroneous handling might open some of the valves unintended, as this may result in a spill.

Consider installing double air filters on harvest tanks and fermentors, and do it in a way which enables the filters to be integrity tested while mounted. It should be possible to fully decontaminate all the way from the fermentor to the second filter without any interference from cold spots.

Search the market for validated tube welders before choosing the type and dimensions of the tubing and connecting devices. An acceptable liquid flow rate must be obtained along with an acceptable way of connecting and disconnecting items, whether for sampling or inoculation. Not all tube dimensions can be welded, nor have all tube welders been validated to an acceptable GMP level.

Cooling water must be protected against contamination in the containment zone. The liquid could either be a disinfectant, a dry cooler, or a double walled system.

In regard to using liquid nitrogen in the production, we found it practical to refill the Dewar buckets through the wall, thereby avoiding having to bring the equipment in and out of the containment zone, which would require autoclave treatment or decontamination. This device, however, must be designed very carefully, as a small amount of liquid nitrogen will expand drastically when it reaches room temperature, if left in the pipe between the valves.

If replacing a component means shutting down the production, a higher priority should be put on the lifetime of the component and its intended use rather than on other aspects such as costs. It is a very complex and expensive task to replace something outside of the regular maintenance schedule, as the production will have to be shut down and decontaminated before any of the systems can be accessed. However, if replacing an item does not interfere with the current production, you might want to use cheap items such as keyboards, which can be replaced and thrown away without any significant costs instead of going for a high IP (internal protection) factor - but ONLY then. Redundant equipment, for example pumps, will make maintenance a lot easier as the production can continue during maintenance periods and the "down-time" of the facility can be minimized during equipment failure.

Ergonomic aspects are an important issue that must be considered, though some aspects might be next to impossible to implement in a containment production. Can aerosol splashes hide inside the construction of height adjustable tables? Some designs are more difficult to keep clean than others.

Should control panels be divided into separate sections for GMP, non-GMP, and containment? Should the control panels for the containment equipment be placed inside the containment zone or together with the rest of the panels in the basement? Is it even possible to install and seal pneumatic and electric installations within the zone in an acceptable way?

Kill System

Designing a kill system and validating it afterwards is a complex task in itself, but designing a kill system for a clean room production facility adds new aspects to be considered. Again, it is important to focus on the clean room aspect.

The volume of waste from a pharmaceutical production is huge, and resembles the amount coming from an animal facility more than anything from an analytical or research laboratory. Fortunately, waste from a pharmaceutical production does not include a significant amount of solids; this makes stirring and heat distribution less problematic than for an animal facility.

80% of the volume going into the kill system might be growth media, which will sustain any microorganisms growing in the tanks and pipes. Designing the system with a buffer or holding tank before the heat treatment tanks, should therefore be avoided, as this will be a good place for a microbiological contamination to start, which might later grow up through the piping and ultimately contaminate the production area. The piping leading from the production area to the kill system should be maintained, sanitized, and steamed regularly — the exact same treatment that the media system is subjected to.

Newly produced waste must be handled immediately in the primary heating tanks, as there is no buffer tank to collect it. As stated earlier, the waste generated by a pharmaceutical production is immense and the tank volumes should therefore be considered carefully.

For safety, a buffer tank can be installed behind the primary tanks. This tank can handle a possible overflow from the continuous production of waste during heat treatment in the specific kill tank. In addition to the growth media, the waste going into the kill system may also consist of eluents from chromatography columns along with CIP fluids and steam/condensate from the process of sanitizing and steaming tanks and pipes. To avoid possible interference such as pressure build-up or backflow of steam and other fluids, it might be wise to lead the steam into the kill system by separate pipes.

It should be possible to decontaminate not only the tanks themselves, but also the whole system, including filters, collecting manifolds, and the pipes leading to them. If the equipment installed around the tanks is not easily decontaminated, an alternative is to wear PPE during maintenance of the open system.

As when addressing a normal SIP procedure in a GMP production, it is relevant to look for cold spots during the commissioning or the validation phase.

Should the system dump waste automatically or should there be a specific batch release on each batch?

Is a continuous decontamination system reliable enough? A risk assessment will help in answering these questions.

What kind of data collection is needed to convince the authorities that the facility and the kill systems are in control?

Is pH adjustment of the waste needed? Monitoring devices should be calibrated inline or decontaminated before being removed from the containment zone for calibration.

Will chlorine ions be present in the waste - perhaps from the column purification steps? This is a very corrosive environment for anything made of almost any type of stainless steel. Are glass coated steel tanks or perhaps very thick black iron tanks an option? Iron tanks also corrode, but slowly and the cost of replacing a tank is fairly low, if the facility has been prepared to allow large equipment to be installed or replaced at a later time. Will the pH electrodes survive the heat treatment?

How is waste transferred back into the kill system once on the floor after an accident?

Is there an airlock or an anteroom in front of the kill room? Are there showering facilities in there? Is there space enough for additional PPE? Where does the water from the shower go? What about towels and contaminated clothes? Which ventilation system serves the area?

Redundancy/Back-Up

Biosafety cabinets should always be on emergency power. Even short power failures may result in a reversed flow, thereby dragging contaminated air into rooms where the ventilation system on emergency power is still running. Freezers should be monitored continuously with alarms in place. Freezers should also be on emergency power, as they will thaw fairly fast, and broken vials of virus working banks will be able to flow out. Pumps, heating mantles, light, and process monitoring computer screens also need back-up.

What should be on UPS (Uninterruptible Power Supply) and what should be on a mechanical power generator? Only a risk assessment can give the answer.

Process Monitoring Systems

Should the system log off the user automatically or should the user himself log off?

Is using initials to accept actions acceptable? Or is a password needed for this? Should the computer or equipment automatically remind the user to change the password? If a wrong password is used, what should be the message on the screen? The words "System failure" are not very informative and it does not inform the user what he or she is doing wrong. Is 21CFR11 (FDA´s regulations for handling electronic data) relevant for the production?

Should it be possible to generate and print reports and alarm logs in the middle of a batch? Or is it acceptable to wait until the batch has been completed? The batch time for a biological process is often close to a month. A lot of alarms on water, steam, and ventilation can be generated during that time.

What happens when the staff wants to print a report containing data from 100 batches? Does it take longer to print this report than one for just one batch? How long is it acceptable to wait for a printed report? Should it include graphs? Test how long it takes to print a report with the graph generation module activated - generating graphs can slow down the report-processing considerably! What about trend data for the utilities? These trend reports include data from several batches. If you need access to your data later, it is very important how these data are stored on the hard disk. This is quite an expensive feature to reprogram at a later stage, so consider it early on in the project.

Are summer and winter time a relevant issue for the facility? Is the system capable of handling a one-hour jump twice a year? What happens to the process data when the hour is changed, either forward or backwards? How much validation will this take? Will it be easier to use winter time or "facility-time" all year around? Create other systems to help remind the staff that they have to leave to pick up their children from the kindergarten before they think they do (because the "facility-time" does not follow the rest of the country for about 6 months).

What level of redundancy is needed for the process monitoring system? What kinds of back-up procedures have you planned for? Must servers be placed inside the containment area? Can they be installed outside the area, perhaps with the use of extenders? Should there be redundant servers? Should these be placed in separate rooms? Is an automatic fire extinguishing system around the servers preferred? How tight must the room be for this? For servers placed within the containment zone, the automatic fire extinguishers might influence the negative pressure or clog the HEPA filters. This depends on which extinguishing principle is chosen.

Address the aspect of color blindness! A surprisingly large percentage of the male population cannot distinguish between red and green. A lesser percentage has problems with other color combinations. Does the alarm status only show on the screen by changing the color of the item? Or does it start blinking? If blinking, a lot of alarms generated at the same time during a power failure will create a screen image that is horrible to work with when starting the facility up again. We tested initial screen images with a selection of confirmed color-blind employees working at the Institute. It was amazing to listen to the discussions that took place between the individuals in the group regarding what they thought they saw on the SAME screen!

Cleaning

It is important to remember the basic differences between a normal pharmaceutical production and a production involving biologicals.

In a normal pharmaceutical production, it is important to make sure there is no residue left on the equipment. The acceptable level of cleaning is determined during validation.

In a production involving biologicals, the equipment must be decontaminated before it is cleaned and reused. The decontamination must focus on killing all viable material, as it is important to make sure that no one who might come in contact with the equipment at a later stage, will be exposed to something harmful, as they might not be aware of the potential risk at that time. After decontamination, the normal CIP and SIP procedures are performed. Being biologicals, these contaminants might multiply and even the tiniest amount left over from a CIP procedure may pose a potential risk at a later time, perhaps by suddenly occurring in much larger quantities than expected.

Decontamination procedures might include autoclaving at 121°C for a specified time period; half an hour at 121°C kills most bacteria and viruses. The process should be validated. This procedure might, however, pose a problem for the next step, the CIP procedure, as a lot of the material inside the vessels might be burned to the sides after having been subjected to the decontamination treatment.

Decontamination of equipment can be done in several ways: heat and chemical treatment are both possible methods. It is important to choose the safest decontaminating method; what is best for the agent, the equipment, the following CIP procedure and of course - the environment.

Documentation

How is the level of containment verified and proved for the containment GMP production? Qualification, commissioning, or validation? Make a risk assessment and focus the efforts on the important issues raised by this assessment.

The amount of extra documentation needed when working in BSL-3 or BSL-4 environments is of course noticeable. However, compared to the demand for documentation already present in a GMP production, the extra documentation seems limited. Most documentation systems already exists scheduled preventive maintenance, policies, SOPs, production records, logbooks, trend analyses, hygienic monitoring, qualification, commissioning, validation, revalidation procedures, DQ (design qualification), IQ (installation qualification), FAT (factory acceptance test), SAT (site acceptance test), OQ (operational qualification), and PQ (performance qualification), just to mention a few. Furthermore, the staff is already GMP- trained and they know how essential it is to document every small observation, no matter how tiny it might seem initially.

It should not be a problem to incorporate validation, revalidation, and batch documentation of new procedures like successful decontamination into the extensive paper system that already exists in regard to batch records, including release procedures and yearly trend reports.

In other words, what should be done is to look at the safety and containment aspects with "GMP eyes" and commission, validate, and document these areas the same way well-known GMP issues would be addressed. Again a risk assessment will point out the areas where validation is needed and where basic commissioning is too low a standard to aim for.

Bottom Line

Biosafety guidelines and GMP rules generally support and reinforce each other. When working with high-risk agents, containment must be achieved - that is: biosafety has the highest priority. However, when working with low-risk agents, the GMP rules can be given a higher priority than the BSL guidelines.

The cost of constructing a GMP facility is generally higher than for a BSL-3 facility. Building a combined GMP and biocontainment facility does not necessarily mean that the budget has to be doubled - especially if the relevant design considerations are taken into account early in the project.

Conclusion

A large amount of guidelines and articles have been written about GMP production facilities for pharmaceuticals, and many guidelines, books, and articles address the biocontainment facilities. Basically, I think that the world of GMP production is very well suited for handling the specific biocontainment procedures and documentation issues, and that most biocontainment focus points have already been addressed in the normal working life of a pharmaceutical producing staff. The extra and specific documentation needed to document compliance with BSL-2/BSL-3/BSL4 is easily incorporated into the already existing GMP system.

Furthermore, it has been a very interesting task to design and build a large clean room production facility while incorporating the future biosafety aspects in every detail. Today we believe that it would have been extremely difficult, maybe even impossible, to redesign an existing normal-to-large scale GMP/BSL-2 facility to comply with BSL-3 or BSL4 requirements.

It was possible to implement most of the containment and GMP aspects in our new building, but still, not only the biocontainment and environmental authorities but also the OSHA and the FDA will be presented with some "balanced solutions" and "compromises" that they are not used to.

It is very important to write down all the aspects and rationales, which form the basis of the design decisions. If the risk assessment finds one set of guidelines more imperative, relevant and important than another, it must be clear why. Risk assessments should be the main tool to weigh out contradicting recommendations seen from a product, biocontainment, environmental, and personal safety point of view. That is the easy part. The difficult task is finding the time to do so and finish it off by writing it all down in a way that makes it possible for you or your successor to simply take these design rationales out of a binder, and remember why choices were made the way they were, and explain this to visiting inspectors in the future.

Build mock-ups of all details, minor as well as major. It is necessary to agree on any specific detail, including performance tests and acceptance criteria, before these details are implemented throughout the building.

As an end-user it is important to join the design group from the start of the project. Do not accept any postulate from the engineers or architects that they can do this on their own. Participate in the building meetings with the contractors during the building phase. Be visible during the commissioning and validation phase, at meetings and during tests on site. By staying away, you have only yourself to thank for working in a facility that does not function according to your expectations or with a suboptimal design. Some contractors and engineers might act a little surprised initially - but they will get used to having you around.

Further Reading: Biosafety

Agriculture & Agri-Food Canada Publication No. 1921/E. (1996). *Containment standards for veterinary facilities*, 1st ed. ("The Red Book").

Arbejdstilsynets bekendtgørelse nr. 864 af 10. (November 1993). Bekendtgørelse om biologiske agenser og arbejdsmiljø.

Arbejdstilsynets bekendtgørelse nr. 177 af 23. (marts 1998): Bekendtgørelse om ændring af bekendtgørelse om biologiske agenser og arbejdsmiljø.

Arbejdstilsynets bekendtgørelse nr.998 af 16. (december 1997): Bekendtgørelse om ændring af bekendtgørelse om biologiske agenser og arbejdsmiljø.

ASHRAE Handbook. (1999). Heating, ventilating and air-conditioning applications.

Bekendtgørelse nr 384 af 26. (maj 2000). Bekendtgørelse om genteknologi og arbejdsmiljø.

Boss, M.J. & Day, D.W. (Eds.) (2003). Biological risk engineering handbook: infection control and decontamination. CRC Press.

Crane, J.T. & Richmond, J.Y. (2000). Design of biomedical laboratory facilities. In D.O. Fleming, D.O. & D. Hunt. (Eds). *Biological Safety: Principles and Practices*, 3rd ed., pp. 283-312). Washington, DC: ASM Press.

DS/EN 1620. Biotechnology - Large-scale Process and Production – Plant Building according to the degree of Hazard.

DS/EN 12128. Biotechnology - Laboratories for Research, Development and Analysis – Containment levels of Microbiology Laboratories, Areas of risk, Localities and Physical Safety Requirements.

DS/EN 12307. Biotechnology - Large-scale Process and Production – Guidance for Good practice, Procedures, Training and Control for personnel.

DS/EN 12460. Biotechnology - Large-scale Process and Production – Guidance on Equipment Selection and Installation in Accordance with the Biological Risk.

DS/EN 12461. Biotechnology - Large Scale Processes and Production – Guidance for the handling, inactivation and testing of waste.

Guidelines for research involving recombinant DNA Molecules (NIH Guidelines), (1998).

Health Canada. (2004). The laboratory bio safety guidelines, 3rd ed. ("The Peach Book").

Lieberman, D.F. (Ed). (1995). Biohazards Management Handbook, 2nd ed. Boston: Boston University.

Richmond, J.Y. (Ed). (1999). *Anthology of biosafety.* I: Perspectives on laboratory Design. Mundelein, IL: ASBA.

Richmond, J.Y. (Ed). (2000). *Anthology of biosafety.* II: Facility design considerations. Mundelein, IL: ABSA.

Richmond, J.Y. (Ed). (2000). *Anthology of biosafety.* III: Application of principles. Mundelein, IL: ABSA.

Richmond, J.Y. (Ed). (2002). *Anthology of biosafety.* V: BSL-4 laboratories. Mundelein, IL: ABSA.

Richmond, J.Y. (Ed). (2004). *Anthology of biosafety.* VII: Biosafety level 3. Mundelein, IL: ABSA.

Richmond, J.Y. & McKinney, R.W. (Eds). (1999). *Biosafety in microbiological and biomedical laboratories*, Public Health Service, 4th Ed. U.S. Department of Health and Human Services, Government Printing Office, Washington, DC

Richmond, J.Y. & McKinney, R.W. (Eds) (2000). *Primary containment for biohazards: Selection, installation and use of biological safety cabinets*, 2nd Ed. U.S. Department of Health and Human Services, Government Printing Office, Washington, DC

U.S. Dept of Health and Human Services. (1999). *Biosafety in Microbiological and Biomedical Laboratories*, 4th Ed. Bethesda: CDC and NIH.

WHO. (1993). *Laboratory biosafety manual,* 2nd Ed. Geneva Switzerland: World Health Organization.

WHO. (1995). Bio safety guidelines for personnel engaged in the production of vaccines and biological products for medical use. Geneva Switzerland: World Health Organization.

WHO. (1999). Global action plan for laboratory containment of wild polioviruses. Geneva Switzerland: World Health Organization.

WHO. (2003). Guidelines for the safe production and quality control of IPV manufactured from wild polio viruses. WHO Technical report, in press. Geneva Switzerland: World Health Organization.

Further Reading: GMP

American Society of Mechanical Engineers. (1997). Bio processing equipment. New York: ASME.

Bygningsreglement. (1995). fra Bygge-og Boligstyrelsen.

Cole, G.C. (1990). Pharmaceutaical production facilities. Design and applications, 2nd ed. Ellis Norwood.

DS/ENV 1631. Clean room technology - Design, Construction and operation of Clean rooms and Clean Air devices. Brussels: European Commission for Standardization.

DS/EN 29001 Quality Systems - Model for Quality Assurance in Design/Development, Production, Installation and Servicing. Brussels: European Commission for Standardization.

DS/EN 29002 Quality Systems - Model for Quality Assurance in Production and Installation. Brussels: European Commission for Standardization.

ISPE Baseline Pharmaceutical Engineering Guide. (1996). Volume 1: Bulk pharmaceutical chemicals. Tampa: ISPE.

ISPE Baseline Pharmaceutical Engineering Guide. (1999). Volume 3: Sterile manufacturing facilities. Tampa: ISPE.

ISPE Baseline Pharmaceutical Engineering Guide. (2001). Volume 4: Water and steam systems. Tampa: ISPE.

ISPE Baseline Pharmaceutical Engineering Guide. (2001). Volume 5: Commissioning and qualification. Tampa: ISPE.

Ljungqvist, B., & Reinmüller, B. (1996). Clean room design: Minimizing contamination through proper design. Buffalo Grove: Interpharm Press. Inc.

NFPA 318. (1998). *Standard for the protection of clean rooms.* Quincy, MA: NFPA.

Odum, J. N. (1997). *Sterile product facility design and project management.* Interpharm Press Inc.

The Rules Governing Medicinal Products in the European Community, Volume IV: Good Manufacturing practice for Medicinal Products, Brussels.

Whyte, W. (Ed). (1991). *Cleanroom design.* Chichester, England: John Wiley and Sons.

Chapter 5

Animal Room Design Issues in High Containment

Leslie Gartner, AIA and Christopher Kiley, PE

Abstract

This Chapter discusses the principals of High Containment Design regarding the Containment Barrier System and Engineering Concepts and Technologies. These principals will be utilized to discuss the issues pertaining to Animal Room Design. High Containment animal facilities contain areas designated as Biosafety Level 4, Enhanced Biosafety Level 3, and Biosafety Level 3 Ag.

High Containment Design

Biosafety Design

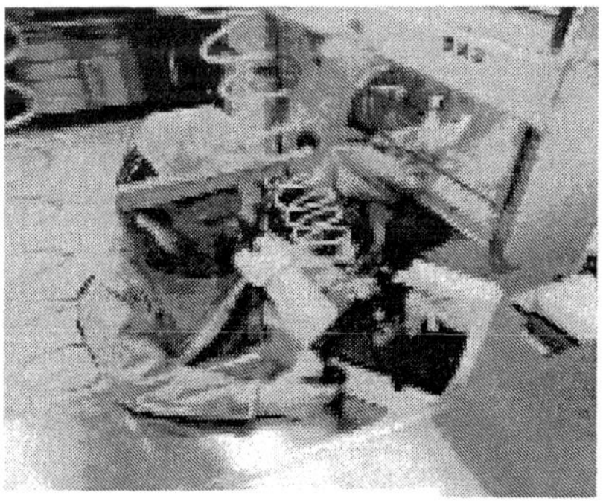

Figure 1. BSL-4 Worker

Most procedures in the laboratories and animal rooms are conducted in primary containment equipment to minimize aerosol exposures to the work environment. Lab work is conducted in a biosafety cabinet, animals are housed in isolators, necropsy and animal procedures are conducted on downdraft / back draft tables, and centrifuge work is conducted with closed routers. The containment barrier is the secondary barrier and is essential for maximum protection of the environment for the procedures that cannot be contained within primary containment equipment. Personnel are protected in biosafety suits that have their own breathing air and HEPA filtration protected exhaust.

All material, animals, air and liquid effluent is decontaminated as it exits the laboratory or animal room. All personnel exiting a BSL-4 suit area are decontaminated in a chemical shower. All services that enter the containment zone have back flow preventors to eliminate the possibility of contamination if the system is temporarily inactive.

Figure 2. Animal Holding Room

The barrier for the BSL-4 labs and animal rooms starts at the clean side of suit chemical shower. The upstream spaces “cleaner” than this room can be shared for BSL-3 and BSL-4 pre-functions (i.e. locker rooms, body shower, and dirty change room). The barrier for the BSL-3Ag labs and animal rooms is located on the clean side of the body shower.

Due to the low additional cost and the advantage for flexibility and surge capacity, the BSL-3Ag laboratories are usually designed to allow use at BSL-4, if the facility includes BSL-4 labs.

Containment Barrier System

The containment barrier is required to provide a gas tight barrier between each independent zone. The *"box-in-a-box"* principal is utilized to ensure the pressure between the containment zones are not influenced by exterior influences of wind and temperature on pressure. Each laboratory, chemical shower, fumigation room and animal room are separate containment zones to allow each to function independently, regardless if an adjacent zone is active or decommissioned for establishing a new program or operating at a different level of containment. The principal of *"pressure differentials"* between the containment zones is employed to create an airlock entry and directional airflow for zones from the chemical shower to laboratories and animal rooms. The leak-tight integrity of the containment zones is crucial due to the immeasurable impacts created on the zones due to adjacent zone activities and environmental conditions.

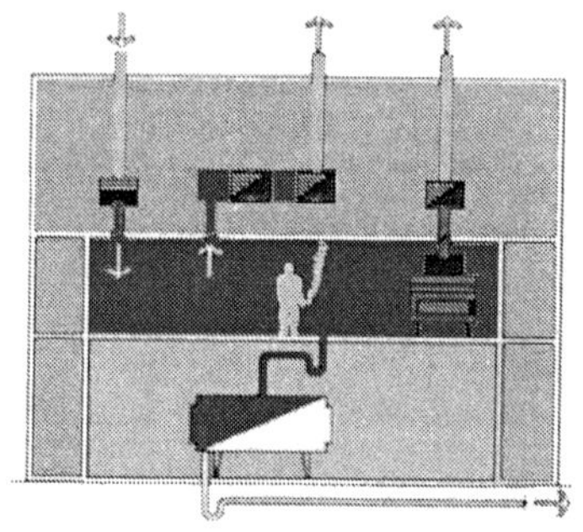

Figure 3. Box-in-box

The containment zones are *"pressure decay tested"* to ensure the gas tight integrity of the containment barrier has been achieved. The pressure decay verification test is required by USDA allowing the facility to conduct research on any Select Agent and meets the requirements of the regulatory agencies. This test provides a constructability standard of acceptance for BSL-4 and BSL-3Ag facilities.

Containment Barrier Components

Personnel and material is required to pass through the containment barrier. The key components and their installation are individually tested to ensure they meet the pressure decay test.

- Air-pressure-resistant doors, allowing entry and exiting, are interlocked to ensure the building system can adjust as an individual sequences in and out of a containment zone.
- Air-pressure-resistant windows allow light in to create good work environments.
- Pass-through autoclaves of various sizes allow for decontamination of total racks and bags of infectious waste. The autoclave, along with fumigation rooms, serves as a pass through for material and animals when the autoclave is clean.
- Dunk tanks filled with decontamination fluid allow for the safe removal of packaged material for further work in adjacent or other labs.

- Supply and exhaust air is HEPA filtered. The HEPA filter housing and contaminated duct work is pressure decay tested to a higher level. The amount of contaminated duct work is minimized by locating the HEPA filter assembly above the ceiling sealed penetration.

- Liquid effluent is collected and treated in the biowaste system via floor and sink drains. The floor drains are capped to prevent overflow of the biowaste system.

- Lab services, including hot and cold water, CO_2 gas, and liquid nitrogen are surface mounted and have special ceiling penetration details with spare capacity.

- Electrical service to the area is from a redundant electrical system backed by standby generators and an Uninterruptible Power System (UPS).

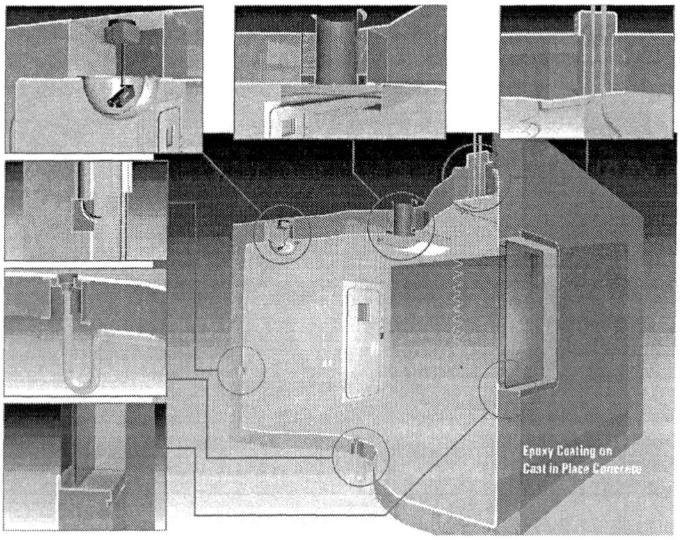

Figure 4. Containment Barrier

- Light fixtures should be recessed in the containment barrier with maintenance access from the mechanical service space above.

- Electrical and data outlets are cast in the containment barrier and sealed with epoxy.

- The containment barrier is covered with a special epoxy coating sealing the penetrations to the barrier construction.

Containment Barrier Construction

The construction strategies and techniques to form a certifiable biological barrier utilize proven materials and methods. Considerations of local construction practices, availability of materials, cost, and past performance are essential variables to be incorporated. Barrier construction, penetration detailing, jointing, sealing and surface applications is crucial to the success of the containment zone being able to meet the pressure decay verification test.

Epoxy on cast-in-place concrete has been selected in numerous facilities, as it is an economical system incorporating known and proven technology, materials and labor techniques. The system has passed the required pressure decay tests on initial start up and on an annual re-certification, which demonstrates a robust and long-term durable design. The surface finish has excellent resistance to chemical resistance and is compatible with other materials (metals) for continuity of the barrier. The robust nature of the design has an inherent seismic sturdiness, compatible with the containment barrier and physical security.

Figure 5. Penetration Box

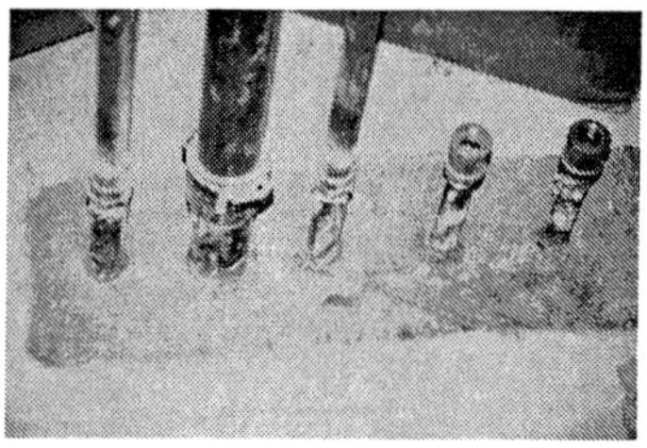

Figure 6. Cast in Penetration

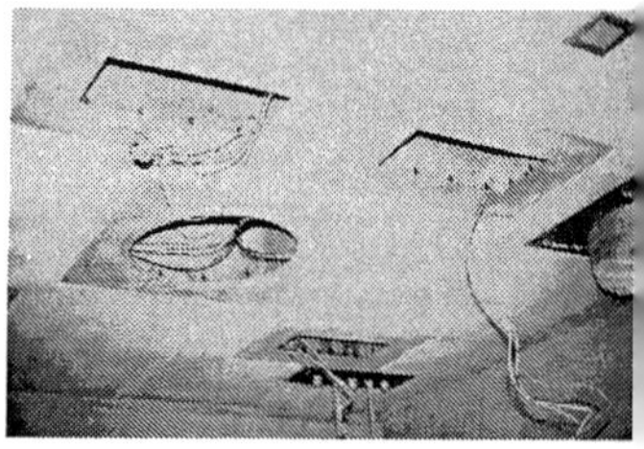

Figure 7. Ceiling Penetrations

The containment barrier epoxy special coating is independently tested against 5 key criteria:

1. Gas permeability

2. Chemical resistance

3. Adhesion to back up material – a pull test is conducted

4. Top coat finish for deconability

5. Recoat capability

The epoxy coating is installed on a cast-in-place concrete structure. Due to the brittle nature of epoxy, the barrier is constructed of poured-in-place concrete. The concrete back-up must be stable to prevent transfer of movement and cracking through to the coatings. The concrete is a low shrinkage mixture, usually utilizing plasticizers to reduce the amount of water and to ensure the concrete reaches all areas. Six mix designs are tested for shrinkage and stability. Three are selected to construct a mockup to test the finish, cracking and stability, and the ability of the mix to seal around all the penetrations. The concrete is specified to cure for an extended period of time until sufficient hydration and maximum shrinkage have occurred. This time is generally four to six months. Application of special coatings should not begin until the majority of curing is complete.

Prior to receiving commissioning approval, concrete surfaces must meet the following criteria:

1. Uniformity of texture

2. Smooth surface finish

3. Acceptable hardness

4. A minimum number of symmetrically spaced form seams

5. All honeycombed areas and tie holes should be filled

6. All projections should be removed to a level flush with surface.

7. Corners and surfaces should be trued-up.

8. Porous or defective concrete surfaces should be repaired by chipping and sandblasting.

9. Patching should be done using only a manufacturer-approved epoxy bonding agent and crack filler.

All components that penetrate the barrier should have a welded steel frame, cast into the walls, ceiling and floor slab. The epoxy coating should be installed over the penetrations and the concrete creating a seamless gastight enclosure.

MEP Concepts and Technology

Mechanical Concepts and Technology

The predominant mechanical methods used for containment are pressure differential control or directional air flow control. Recent BSL-4 containment facilities utilize both control methods in a two stage process. The supply and exhaust air quantities are controlled to a set volume (typically based on required exhaust volumes) and then the supply air is controlled or "trimmed" to maintain a room pressure set point. The devices used for the room pressure trimming would be best located on the "clean" side of any gas-tight dampers and HEPA filters in order to facilitate gas decontamination.

Utilizing a Class III glove-box for containment purposes makes the construction of the architectural barrier much simpler, as well as the mechanical system. Disconnecting the supply (not hard ducted) from the glove-box and HEPA filtering the supply inlet to the box (es) will prevent the possibility of over-pressurizing the glove-box. Sufficient exhaust quantities need to be calculated in order to achieve the static pressure required for the glove-box to maintain containment in the event of a glove or multiple glove breach.

Containment Biosafety Matrix

Lab Type	Chemical Shower	Body Shower	Pressure Decay Test	Gas Decon	Double Exhaust HEPA Filter	Single HEPA Exhaust Filter	HEPA Supply Filter	Effluent Decon	Breathing Air	Pass thru Autoclave
BSL-4 Laboratories	Yes	Yes	Yes	Yes	Yes		Yes	Yes	Yes	Yes
BSL-3 Ag Laboratories		Yes	Yes	Yes	Yes		Yes	Yes		Yes
Add to BSL-3 Ag Labs for use at BSL-4	Yes								Yes	
BSL-3 Laboratories		Yes *				Yes *				Yes
BSL2 Laboratories and Lab Space										
A-BSL-4 Animal Space	Yes	Yes	Yes	Yes	Yes		Yes	Yes	Yes	Yes
BSL-3Ag Animal Space		Yes	Yes	Yes	Yes		Yes	Yes		Yes
Add to BSL-3 Ag Space for use at BSL-4	Yes								Yes	
A-BSL-3 Animal Space		Yes		Yes *		Yes		Yes *		Yes
A-BSL2 Animal Space		Yes								
ACL-3/4 Insectory Space	Yes	Yes	Yes	Yes	Yes		Yes	Yes	Yes	Yes
ACL-3 Insectory Space		Yes		Yes *		Yes		Yes *		Yes
ACL-2 Insectory Space		Yes								Yes

* Program- and agent-specific requirement

The following conceptual section demonstrates the potential servicing concepts associated with a High Containment layout.

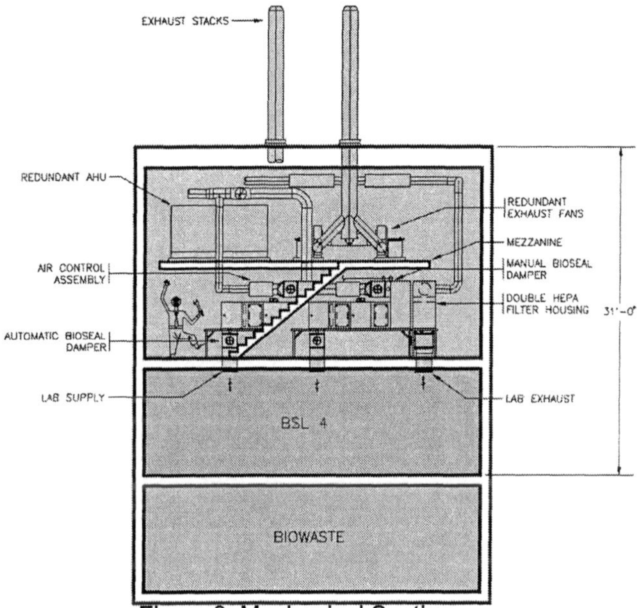

Figure 8. Mechanical Section

- The HEPA filter level is located directly above the High Containment Suites.
- The HVAC units are located on a mezzanine directly above the HEPA filters.
- The Biowaste Level is located directly below the High Containment Suites.

Electrical Concepts and Technology

The most common electrical issues related to high containment laboratories are standby electrical emergency power or UPS systems and redundancy. The issue is not only to what extent emergency standby power should be provided, but which systems or equipment cannot tolerate downtime during generator start-up and, consequently, must be fed by a UPS system; also which systems require 2N or N+1 levels of redundancy from a local and building/ campus standpoint. Typically all panel boards are located outside of the containment zone.

Provisions for the mounting and/ or servicing of light fixtures have two concepts. One concept, which was utilized in the Winnipeg facility, is to cast a lensed-box into the ceiling and mount the light outside of containment in the interstitial space. A second concept utilizes surface mounted fixtures in containment with minimal penetrations made to the barrier.

The increased use of electronic equipment, including computers and facsimile machines, for the transfer of data across the containment barrier is prevalent in all the facilities studied.

Plumbing Concepts and Technology

Biowaste is treated through heat decontamination in all of the facilities benchmarked, although the various methods differed with respect to system assemblies. Several systems used holding tanks prior to loading the heat/ pressure vessel(s); this application would be more cost effective as the pressure vessels are a large expense. Piping from these systems ranges from welded stainless steel to polypropylene. A study should be made to the appropriate piping selected based on compatibility of fittings between the bio-waste sources and the final liquid effluent decontamination tanks, particularly with respect to the effects of steam.

Figure 9. Biowaste Tanks

Fire Protection

The issue of fire suppression in high containment laboratories requires consideration of the following issues:

- Life Safety (fire control and containment; wall, floor, and ceiling fire separations; low occupancy)
- Biosafety (how to clean up contaminated discharge; reliability of system; pressurization of the lab)
- Combustibility (minimal combustible material and durable finishes)

It is anticipated that high containment areas should be separated from other areas of the building by fire separation walls with limited, secured occupant access.

A wet sprinkler system with protected heads provides the most reliable system which minimizes accidental releases due to electronic failures or physical damage. The bio-waste system provides a methodology to decontaminate the water discharge. A holding tank will provide adequate water to deploy sprinklers in the largest space and will limit the total outflow of water to reduce the risk of over extending the capacity of the biowaste system.

Bio-Security

Security of the high containment facilities addresses protection of the facility and its occupants as well as protection or integrity of the agents maintained. The specific issues of high containment facilities are that small samples are to be protected, there are no electronic tag and detection devices that can be employed and these facilities function on a collaborative research approach, which necessitates sharing of information.

The bio-security program includes security management, security plan development, risk analysis and countermeasure, which includes physical security solutions.

Physical Security should employ a layered approach to control access of personnel and materials to and from the facility.

- Perimeter Security – Property Protection Area
- ▫ Physical barriers: fences
- ▫ Monitoring and detection: cameras, low light level, visual observation
- ▫ Access Control: staff, visitors, deliveries

- Building Security
- ▫ Set back from perimeter security: 80 m
- ▫ Building hardening: glazing, blast resistance
- ▫ Access Control: Single point of access? Staff, visitors, deliveries

- Security Personnel and Monitoring - Entry control, screening, cameras
- ▫ Laboratory security
- ▫ Layered access: entrance to building, access to common areas, access to Basic Laboratory zone, access to High Containment zone, access to Animal Containment zone, access to Office zone
- ▫ Single point of access control to High Containment zone: people, materials; how to avoid piggy backing?
- ▫ Layered access control to High Containment zone: change room access, inner change access, and High Containment Suite access.
- ▫ High Containment Suite access: unique identifier such as biometrics
- ▫ Material access: freezers containing stock BSL-3 and 4 materials to be kept with suite
- ▫ Monitoring: security station

Animal Holding Room Design

Much basic infectious diseases research requires the use of laboratory animals to study pathogenesis and host responses. Animal rooms at BSL-4 are generally designed to house multiple species including rodents, rabbits, and nonhuman primates.

Individual studies to study infectious agents of bioterrorism and emerging agents, generally require relatively small numbers of animals (<500, rodents, <16 nonhuman primates) and generally last from about 3 weeks to 3 months. Recent research facilities are primarily intended for pathogenesis research, however, for flexibility, the facilities should be designed to allow portions of its laboratory space for the development and testing of new or improved vaccines and therapeutic agents for the prevention and treatment of diseases caused by potential bioterrorism agents or emerging pathogens. Depending on their experimental design and goals, such studies may require larger numbers of animals than indicated above, and may require longer than 3 months for completion.

Planning and Design Issues

Programming

The *BSL-4 Animal Holding* component of the program consists of containment space to support *in-vivo* protocols with Category A, B, and C agents at BSL-4 safety level, office space to house resources necessary to conduct, maintain and service the protocols, and noncontained support functions necessary to maintain the systems, material supply and waste streams for these studies.

Containment space at BSL-4 containment level may include:

- Animal holding
- Basic and advanced procedure rooms
- Aerosol challenge
- Entry/exit

Resource spaces include the functional office, meeting and administrative requirements for core animal research staff. The functional resource areas may include:

- Core research PIs
- Veterinary/husbandry
- Pathology
- Histopathology
- Aerobiology

Animal support functions may include:

- Cage wash and clean cage processing
- Clean animal holding

Space allocations for animal receiving and waste stream disposal functions are included in the building support component of the program.

Grey-out space for holding animals long-term post exposure is generally not included in the program requirements at high containment, as studies will be conducted under BSL-4 containment, where it is assumed that all animals will be sacrificed and undergo necropsy.

Telemetric Monitoring

Biomedical telemetric devices are biosensors that detect physiological parameters and convert them to digital form. Commonly monitored physiological parameters include electrocardiographic data, temperature, oxygen saturation, blood pressure and respiration. The digital signal from the biosensor is then output to a transferring device, which for wireless devices means utilizing a radio frequency transmitter and receiver, to a remote unit for display, analysis, and storage.

The use of biomedical telemetric devices is being considered for animal and NHP subject monitoring within research facilities. Telemetric monitoring will provide continuous monitoring of physiological parameters, reducing the need for animal manipulations and limiting the potential for pathogen exposure and cross-contamination.

Incorporation of Telemetric Monitoring within a research facility must consider the following design and planning issues from the outset:

- Sensor Design: The design of telemetric sensors should consider their application to a variety of small animal and NHP subjects. Tamperproof and possibly injectable sensor devices should be considered for NHP applications while common/standard sensor design may be appropriate for other animal subjects such as mice or rabbits.

- Frequency Bandwidths and Interference: With the increasing use of facility-wide wireless systems and the potentially large number of monitored subjects, the demand for bandwidths devoted to telemetry systems will be substantial. The level of demand may require a dedicated radio-frequency spectrum with the capacity to accommodate the bandwidth needs. Furthermore, the spectrum must be free from interference from other transmission sources, such as MRI, in order to avoid corruption of telemetry data.

- Advanced Imaging Technologies: Disease progression can be monitored through the use of advanced imaging devices. MRI and other technologies may conflict with telemetry, not only in terms of interference but in device design and construction. If possible, solid-state, nonmetallic sensors should be considered.

Process Flow

The animal housing component of high containment facilities functions as a central source repository of material and data to be read or extracted and further processed with core technologies. Throughput capacity for animal studies requires detailed investigation to validate assumptions and animal protocol size and duration will vary based on research needs. The degree to which observations and procedures, telemetry and advanced imaging technologies will contribute towards the research endpoints is a key factor in determining layout.

Supply and Waste Movement should occur via dedicated route of material flow between Animal Support and Animal Holding in containment. Material must be sterilized before leaving containment. Material supplies will cross the containment barrier into Contained Animal Holding through:

- Fumigation Rooms
- Double door bulk (floor model full-rack) size autoclaves
- Dunk tanks

Design Criteria

The Animal Holding Rooms are required to be built to containment standards and requirements defined in the BMBL for ABSL-4 Animal Space and/or standards defined in USDA ARS Guidelines. The following are design issues to consider when design high containment animal holding rooms:

- The containment barriers around the animal rooms and fumigation rooms and chemical shower are to be constructed to meet the pressure decay test.
- The high containment block should be designed to remain operational after a seismic event as per International Building Code.
- The interior surfaces of walls, floors, and ceiling areas are constructed for easy cleaning and decontamination. Walls, ceilings, and floors should be smooth, impermeable to liquids and resistant to the chemicals and disinfectants normally used in the laboratory. Floors should be monolithic and slip-resistant.
- A combination of gasketted and pneumatic bioseal Air Pressure Resistant doors that facilitate access to the animal rooms, chemical showers, laboratories, and fumigation rooms should have varying access devices from biometrics to request to enter buttons, depending on the location of the door and access control required.
- Penetrations in floors, walls, and ceiling surfaces are to be cast in and sealed gas tight.
- All exhaust air and supply air should be HEPA filtered. HEPA filter assemblies should be located directly above the penetration in a controlled mechanical space, reducing the amount of contaminated duct work and to allow independent certification and decontamination of the filter.
- Drains positioned in the Animal Holding Room should be of the deep seal type and be filled with chemical disinfectant of demonstrated efficacy against the target agent, and should be directly connected to the liquid biowaste effluent decontamination system. A wash down system should be provided in each animal room.

- Water distribution piping, vent lines, CO_2 gas distribution piping in the necropsy rooms should be surface mounted on stand-offs within exposed recessed vertical service raceways cast in the concrete. These services should be positioned in an orderly manner within close proximity to their final connection points in order to minimize horizontal surface area.

- Breathing air distribution system should be surface mounted with standoffs. An eyewash station and emergency shower should be provided inside the animal room zone, if the rooms are to be utilized as BSL-3/3Ag animal rooms.

- A wet sprinkler pipe system with control of water volume can be provided with each zone having one penetration and shut-off valve.

- Conduit for all systems such as electrical, data/communication, fire alarm, and security should be installed in the concrete containment barrier and filled with epoxy.

- Light fixtures should be either recessed into the containment barrier slab or IP65-rated surface fixtures to allow washdown.

- Programmable central lighting control system and local time switches should be utilized for animal housing lighting control.

- Fixed equipment should include pass-through autoclaves capable of decontaminating a full rack, biosafety cabinets, and back draft tables in procedure rooms, down draft necropsy tables, aerobiology equipment, dunk tanks, and LN_2 storage units.

Nonhuman Primate Animal Holding Rooms

- NHP in Isolator Racks, either in 2 unit module - over/under, or 4-unit module with flexible isolator.
- A minimum of 16 NHPs.
- Dedicated procedure room to control cross-contamination.
- Wall between holding room and procedure room to separate acoustic and visual.
- All equipment mobile for flexibility of layout and ability to utilize for other functions.
- One pressure zone. Each side of the zone has it's own supply/exhaust to allow directional airflow in either direction.

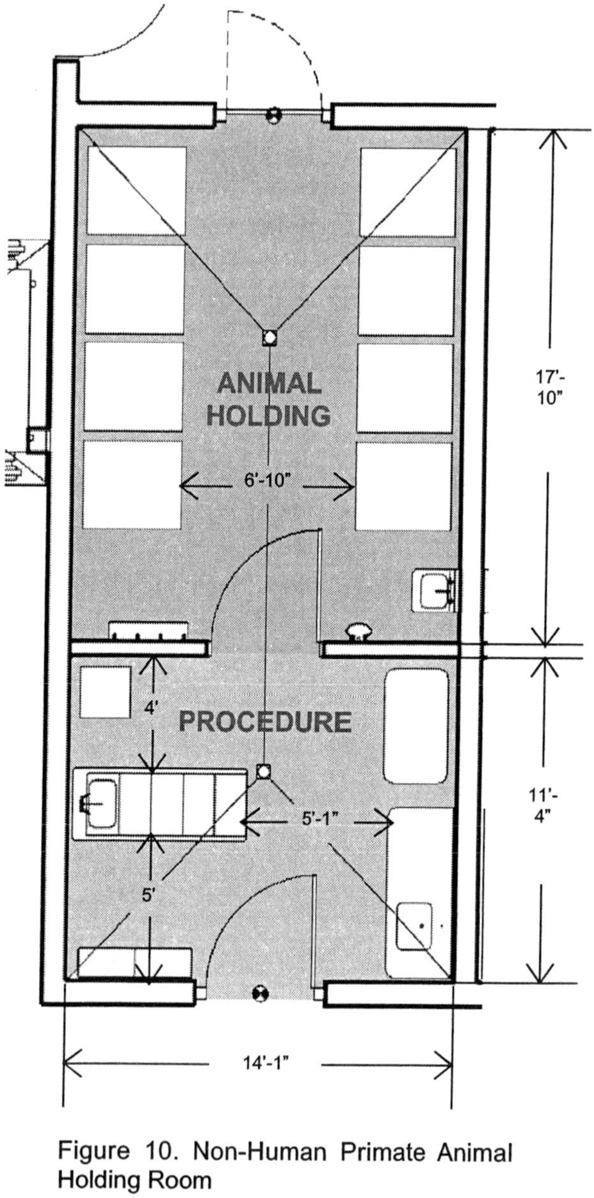

Figure 10. Non-Human Primate Animal Holding Room

Small Animal Holding Rooms

Small Animal Holding Room

- In isolator racks and/or isolator cages.
- Biosafety cabinets for procedures.
- Separate pressure zone.
- All equipment mobile for flexibility of layout and ability to utilize for other functions.

Necropsy Room

- Separate pressure zone.
- Design for 2-person necropsy and one person at tissue station.

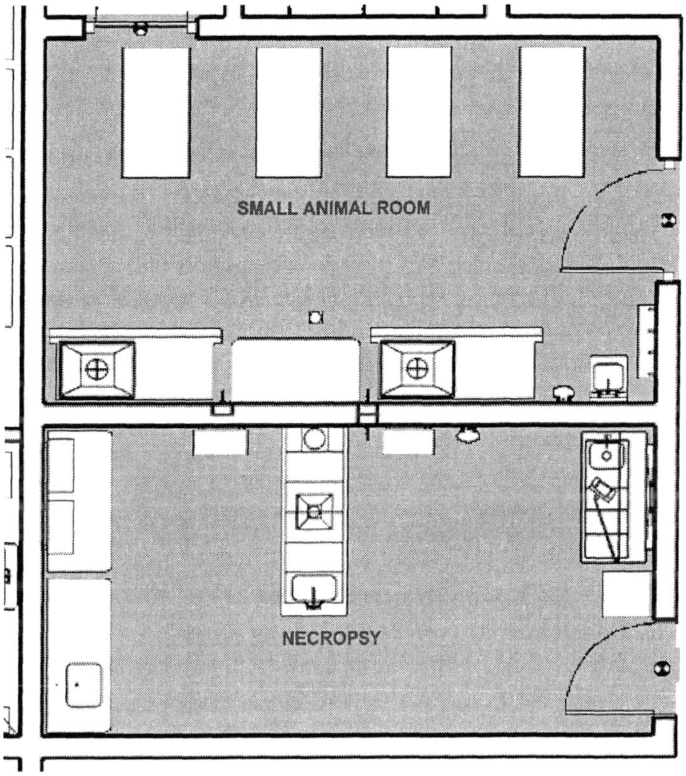

Figure 11. Small Animal Holding Room

Engineering

Animal Rooms Distribution Concepts:

Animal rooms within containment spaces should be designed for maximum flexibility. Due to the nature of the construction techniques in maintaining pressure/air tight spaces it is very difficult and costly to modify the mechanical systems to be individually tailored for the requirements of each animal species that may be used in these spaces. Each animal room should be individually controlled and should be capable of providing 15 air changes per hour (ACH) with the ability to reduce air change rate to match the program needs. The systems should be capable of controlling the spaces in a range of 68°F to 85°F (+/- 2°F) and a relative humidity between 30% and 70% (+/- 5%) (ILAR - *Guide for the Care and Use of Laboratory Animals*). In addition room side replaceable 30% filters should be utilized so that the HEPA filters last longer.

When designing for animal spaces, consideration should be given to the location of the supply and exhaust diffusers to allow even distribution of air as well as temperature and humidity within the spaces. The traditional method has been to supply at the ceiling and exhaust near the floor to create an even distribution of air across the space and to limit "dead" air spots. As can be seen from the CFD analysis (figures 12 through 15), the distribution of air as well as the temperature is the same or better if the supply and exhaust devices are both located near the ceiling. This also adds another benefit in that the amount of contaminated duct is reduced, and there are fewer horizontal surfaces within the space.

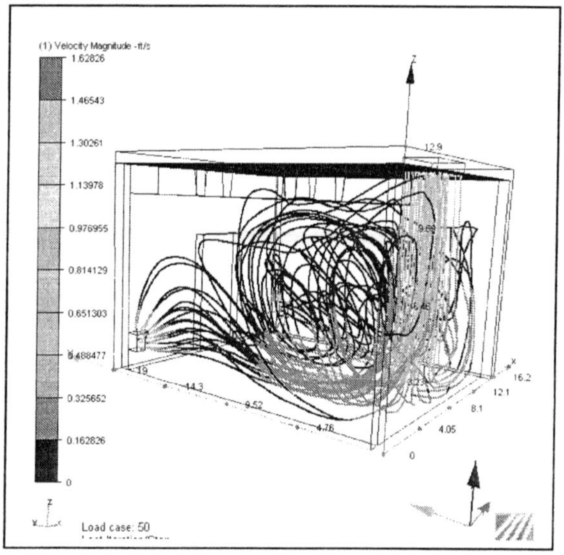

Figure 12. Velocity Profile - Large animal Room with low exhaust.

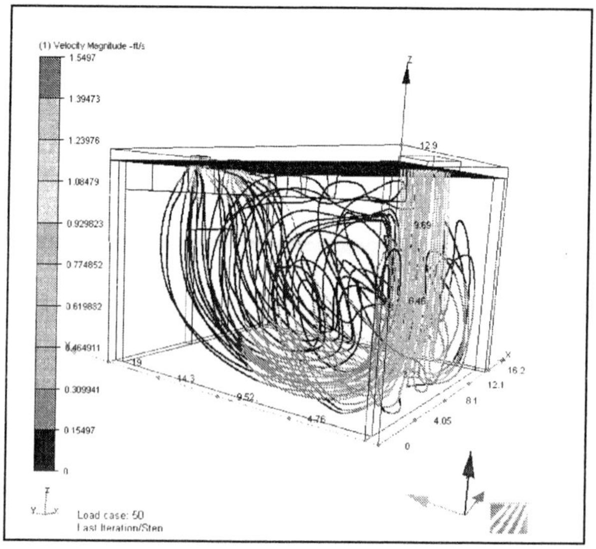

Figure 13. Velocity Profile - Large animal Room with High exhaust.

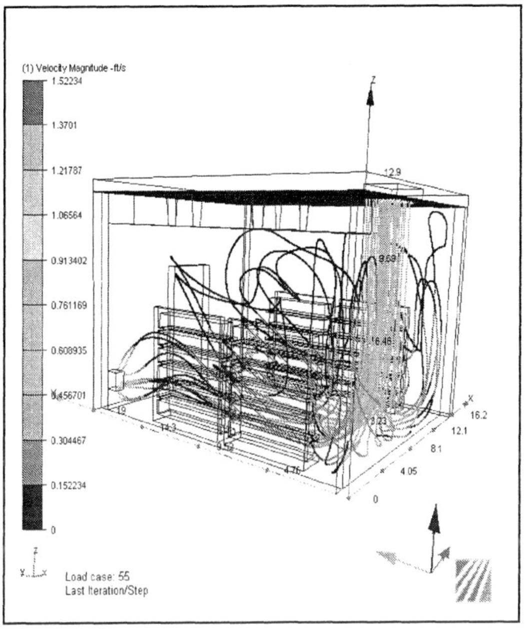

Figure 14. Velocity Profile - Small animal room with low exhaust.

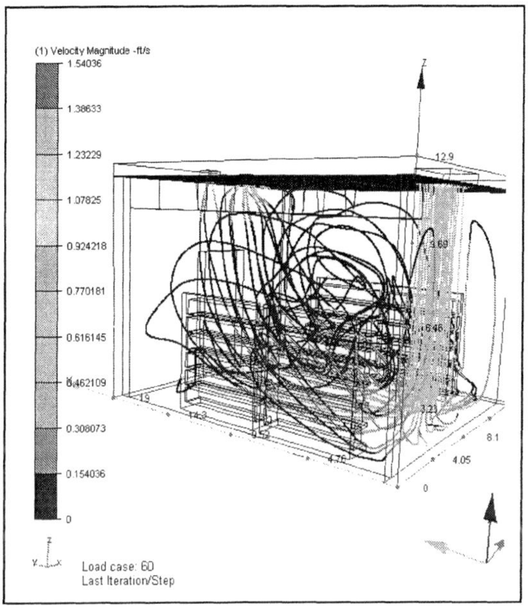

Figure 15. Velocity Profile - Small animal room with low exhaust.

Summary

Animal rooms in high containment need to meet strict biosafety guidelines and remain functional to allow efficient operations. Most small and medium species can be housed in primary containment; however procedures such as removing the animals from their cages, require the animal room to act as the primary containment during these times. The technical challenges of meeting the stringent requirements of designing the room as the primary containment barrier requires the rooms to be designed as an integrated system, involving the barrier, penetrations, mechanical and electrical components, and equipment. The HVAC system and distribution concepts need to be responsive to the functional need to provide flexibility for a variety of species and to provide a reliable dynamic system to control the separate pressure zones for containment.

Acknowledgements

Special Thanks to our colleagues at Hemisphere Engineering, Inc.: Kelly Bistritz, Cory Ziegler, and Harry Wiber.

References

Richmond, J.Y. & R.W. McKinney, (1999). (Eds.) *Biosafety in Microbiological and Biomedical Laboratories.* U.S. Government Printing Office, Washington, DC. 4th Edition.

Institute of Laboratory Animal Resources, Commission on Life Sciences, National Research Council. (1996). *Guide for the Care and Use of Laboratory Animals.* National Academy Press, Washington, DC.

United States Department of Agriculture, Research, Education, and Economics. (July 24, 2002). *ARS Facilities Design Standards – 242.1M-ARS.* Facilities Division, Facilities Engineering Branch.

Chapter 6

Mobile/Modular Containment Facilities

Monica J. Heyl, Charles E. Henry, and Dennis J. Reutter

Introduction

The concept of mobile/modular analytical laboratory facilities capable of performing analysis on-site at the location of suspect samples has come to fruition in the 21st Century. This requirement has long existed; however, not until recently has the need for these capabilities become so apparent that we must fully address the design, development, integration, and validation of mobile and modular analytical laboratories.

Designing and developing laboratories for field application requires the evaluation of state-of-the-art, highly technical, emerging technological trends and the selection of appropriate novel technologies based on a variety of criteria such as engineering, technical risks, and logistical burdens. At the Edgewood Chemical Biological Center (ECBC), several comprehensive (chemistry, biology, radiology, and high-explosive residue) transportable laboratories have been developed for customers with national- and international-level missions that include the execution of the Chemical Weapons Convention (CWC) and Weapons of Mass Destruction (WMD) countermeasures. Hardware adapted for field use, applicable methodology, and customer training are provided, integrating disparate disciplines into comprehensive, turn key packages that can be deployed to remote regions with little or no logistical support and that produce data that are legally defensible.

Along with the requirement to assess unknown or dubious samples of suspected biological, chemical, radiological, and explosives origins, comes the requirement to assess and analyze these materials safely, securely, accurately, and decisively. Thus, this chapter will address the profound requirements, the emerging challenges, and recommend possible solutions that although may be novel, follow prudent practices and established guidelines.

Recognizing the Challenge

Why has the demand for mobile analytics emerged with such gusto? The easiest explanation is terrorism, although potential terrorist threats are far from the only answer. The Defense Department, law enforcement, public health, local communities, regulatory agencies, and commerce as a whole struggle with the burden of waiting for answers. Decisive answers save lives, dollars, and our environment. Unfortunately, decisive answers in today's world must overcome technical engineering barriers and address congruent technologies and disparate disciplines. Take for instance the potential consequences of analyzing dual threat material such as inhalation anthrax and a chemical explosive. Addressing such potential threats poses inherent challenges, but finding the appropriate solution will result in saved lives, not to mention dollars, infrastructure, and our environment. Although frightening and a very real possibility, this scenario is not solely why we develop mobile facilities. Many regulatory decisions made at the federal level have tremendous impact on our lives:

(1) As stated in the first bullet of the Department of Commerce's Mission Statement, we "build for the future and promote U.S. competitiveness in the global marketplace, by strengthening and safeguarding the nation's economic infrastructure".

(2) Likewise, the mission of our Department of Health and Human Services is to serve as the U.S. government's principal agency for protecting the health of all Americans by providing essential human services.

(3) We rely on our American Public Health Association to protect our environmental and community health.

(4) The mission of the Environmental Protection Agency is to protect human health and the environment.

All of these agencies are required to react rapidly to a potential biological, chemical, radiological, or explosive event and gather information that will enable decision makers to make informed decisions in a timely manner to protect our homeland.

The paradigm that exists is that while the need for mobile analytical laboratories has grown, the supporting analytical instrumentation has become smaller, faster, and provides superior performance and more in-depth data, allowing for drastically improved capabilities. This trend will continue and will lead to the continued sophistication of field analyses. However, along with the emergence of more sophisticated field analytics comes the requirement to develop enhanced infrastructures to allow for their use in appropriate environments. And, accordingly there is the need to institute robust quality assurances, appropriate engineering designs and controls, processes, procedures, and reoccurring and institutional training.

Clear guidance addressing engineering controls is not established for mobile or modular laboratory facilities. By leveraging from the recommendations as outlined in the series of Biosafety Anthologies, many practices are available to consult when designing levels of containment to include any one of the four biosafety levels (BSLs) termed BSL 1, 2, 3, and 4, as defined by the CDC/NIH guidelines (Richmond & McKinney, 1999).

The introduction of mixed and unknown hazardous materials that may include any type of chemical, traditional and nontraditional chemical warfare materials, radioactive or nuclear materials, and conventional and unconventional explosives poses considerable challenges in the design of mobile and modular analytical laboratory facilities. Additionally, the urgent needs of our federal and state partners, expedited timelines due to immediate threats, and declining financial resources during times of conflict and terrorism are not good reasons or excuses to cut corners in laboratory design. In fact, they are compelling reasons to approach design safely and security with innovation, cost effectiveness, and efficiency.

Approach

To best address the recognized challenges, an approach has been established to develop mobile/modular laboratories. The approach is simply to:

- Clearly define requirements.
- Establish a peer review committee at the inception of a product.
- Develop the best technical approach.
- Build the facility.
- Verify/Validate.

Requirements

Defining a customer's requirements often is one of the most challenging steps in the process. Understanding and then defining a customer's requirements may be based on factors such as who are the operators, what is their skill level, where will the labs be positioned, what is the annual operation and maintenance budget, and a variety of additional factors. It must be determined specifically what they need and what they want to do. Credibility - quality control - is paramount. Quality control measures are invariant with the location or the type of laboratory. We recognize that whatever needs to be done, needs to be done as well in a mobile/modular facility as in a fixed-site laboratory.

Peer Review

After requirements are clearly refined and agreed upon by the customer and the integrator, we establish a peer review committee. The committee must be comprised of (at a minimum) the customer, user representative, integrator, risk managers for safety, security and environmental quality, appropriate subject matter experts, and the appropriate consultants. Due to inherent risks associated with designing facilities (mobile, modular or semi-fixed) that address overlapping disciplines, this step is extremely important. We bring together a group of experts, consisting of the lead engineer, the program manager, the biologist and/or the chemist, the vendors that we may be working with, and most importantly, experts in health and risk management.

Typically, this team is working to meet or exceed federal or industry standards based on requirements and disciplines. It is important to establish which design criteria the team is working toward: DOD, FDA, NIOSH, CDC, USDA, NIH, or all of the above plus several others.

We pull in the very best experts in the field to make sure that when the time comes for final assembly approval, no one will suddenly reveal that we haven't met a particular standard. Instead, we let the experts tell us in advance what design criteria we will be achieving and how to meet such design criteria so that when our product is ready for delivery, it incorporates all of the customers' and experts' expectations.

Best Technical Approach

As part of our process, we develop a document that is termed a "Best Technical Approach" (BTA). The BTA is based on, but not limited to, the requirement definition, literature searches, gatekeepers (go/no-go criteria), cost, timelines, operation and maintenance budgets, expert advice, and the appropriate available technology. The BTA includes an examination and rationale of the trade-off involved in various approaches. The BTA recommends an approach to the design and development of the system based on the particular customer's needs. It also provides rationale for each recommendation. The BTA makes recommendations; however, the customer makes the decisions on the path forward.

Build

Following the customer's decision on final laboratory design and configuration, we commence all necessary work to fabricate the laboratory. This work shall include procuring platforms, instruments, engineering controls, support equipment, and ancillary items; designing and constructing packaging for selected components and subsystems; and integrating all instruments, components, and related equipment into an operational system.

Verify/Validate

Based on the peer review process, types of validation that are to be performed have been previously decided. Verification is ongoing throughout the project and includes ensuring that each sub-system is working in accordance with the manufacturer's specifications. Validation is a process that ensures the resultant systems work in accordance with the customer's requirements. Validation includes in-house and external review by independent consultants and the appropriate subject matter experts.

Integration of Primary Containment and Analytical Instrumentation Within Engineering Controls

Super Toxic Analytical Glovebox System (STAGS)

Congress has mandated that our stockpiles of chemical weapons be destroyed. The Alternative Technologies to Incineration of Chemical Weapons Program (ERDEC-SP-053, 1996) was established to recommend suitable destruction technologies. This program required the Army to conduct a purity analysis and major impurity identification on the agents HD and VX in bulk ton containers stored at various U.S. Army installations. In that several of our storage sites do not have analytical laboratories, a transportable system capable of analyzing neat, chemical agent materials was required. The solution for this capability resulted in the Super Toxic Analytical Glovebox System

(STAGS) (Patented, 1998). The STAGS is a novel, cost-effective and timely solution to the analysis of lethal materials anywhere in the world.

The design of the STAGS began by initially establishing a working group of specialized personnel at the inception of the concept. The conceptual design team consisted of experts from our in-house team, Army Safety, Industrial Hygiene, and Environmental offices, an engineer who represented the selected analytical instruments (Hewlett-Packard), a former president of the American Glove-box Society, a former president of the American Biological Safely Association, the customer, and a representative from the Alternative Technologies Team.

The STAGS is an integration of robust engineering controls (glovebox) and analytical instruments. The system is designed with two distinct compartments. The analytical portion (Hewlett-Packard GC/MSD/TCD) is enclosed in one of the compartments similar to a small fume hood where HEPA filtered air is drawn through and exhausted. The oven and mass spectrometer air are vented separately through ducts to prevent heat buildup, and a panel with temperature controllers is mounted on the outside of the enclosure so that it is not necessary to open this compartment.

The glove-box portion was welded to the top of the hood and required the integration of several unique features. The double door airlock on the glove box had to be removable. This was necessary so that it could be reversed and reside inside the glove-box during transport. The injection port is an airtight assembly and is maintained at a constant temperature to prevent "cold spot" vapor trapping. It was also necessary to ensure that vapor could not be transferred from the glove-box side of the assembly into the analytical portion through the removable injector assembly. This assembly is temperature controlled, easy to manipulate, and protected so the operator cannot accidentally be burned because of the extremely high temperatures required for vaporization of the samples. The glove port height on the glove-box viewing window was decided by an average height among the chemists from the operations team and it was the determining factor as to the height of the lid that covered the glove-box and control panel assembly. The lid serves as the base of the unit when it is set up for operation and it is equipped with a cutout door at ground level so that the operator can place his or her feet in a comfortable position while working.

Air filtration, the real heart of the system, was a major concern for the safety community. It was necessary to build in redundancy with every major component to ensure a fail-safe system. This consisted of dual High Efficiency Particulate Air (HEPA) and carbon filters, dual vacuum gauges, dual air pressure alarms, dual motor blowers to draw air through the glovebox, dual sources of electricity, and a Flame Photometric Detector (FPD) to monitor between the carbon beds for possible breakthrough of chemical compounds. This was necessary because the effluent air from the filter beds was being released into the working environment of the operators. Army M48 gas and particulate filters were selected because of their size, weight, and predisposition of being approved by the safety community for agent filtration. Their initial drawback was that they were in short supply; however, they did offer a proven, validated filtration system in a short amount of time.

Once the system was designed, blueprinted and approved, contracting and fabrication required approximately 3 months during which all core team members worked very closely. This included determining worst case and what-if scenarios for potential accidents, accident remediation, writing and obtaining approval for operating procedures, and Department of the Army Safety approval for use on Army installations.

Although the application of engineering controls on this project were for chemical agent material analysis, it was quickly obvious that this concept was relevant to infection agents and unknown threats (Henry, Heyl, Reutter, Diez, & Landy, 1997).

Figure 1. Super Toxic Analytical Glove-box System (STAGS)

Primary Containment for Unknown Samples

Customer requirements have led us to a focus on primary containment for unknown samples. This has been an evolutionary process that began approximately 5 years ago.

Primary containment often begins with an *in situ* situation such as the wearing of personal physical protection while collecting and screening samples. Screening of samples, particularly with potential explosive components, is best performed at the site of origin. Once knowledgeable and trained personnel perform a preliminary screen, samples can be moved into more robust primary containment for further assessment.

Field portable, self-contained Class III glovebox type systems, equipped with redundant HEPA, TEDA Carbon filtration have been developed and fielded. These systems are capable of receiving unknown materials including the most toxic of chemical warfare agents, toxins, radiological/nuclear, and infectious biological agents. Systems of this type are typically used for physical examination of unknown samples (limited in size), and then provide a platform for further identification using prescriptive, well-defined protocols and assays for assessment.

The developed system served a particular customer's requirement; however, it has since been patented (Patented, 2003), licensed to industry, and reproduced more than 50 times by our industrial partner. Several objectives were met with the development of the portable glove-box and filtration system. It provides for a portable containment system that can be readily transported and deployed at remote locations. It also provides a containment system of efficient design that includes a self-contained glove-box and a filter system with replaceable filter elements. Further, it provides a modular containment system that can be efficiently adapted for different requirements. These capabilities, along with addressing appropriate protocol development, have proven to provide transportable engineering controls addressing the all-threat sample container concept for receipt of potential NBC materials.

The development of the transportable glove box and filtration system began by assessing established military filtration technology and then transitioning the filtration technology into new platforms. The military technology that was transitioned was found in the M48 filter system. The redesigned filtration system then underwent established military testing to include the Military Specifications for NBC Filtration in Class III Chem/Bio Safety Enclosures,15-Minute Rough Handling for Military Standard 828, Method 105.11, Tested for DMMP Leakage/Life IAW Mil-F-51525, Class III Glovebox, Exhaust Filtration System, w/75CFM HEPA & ASZM-TEDA Carbon Filtration (Military Specifications for NBC). (See Figures 2 and 3).

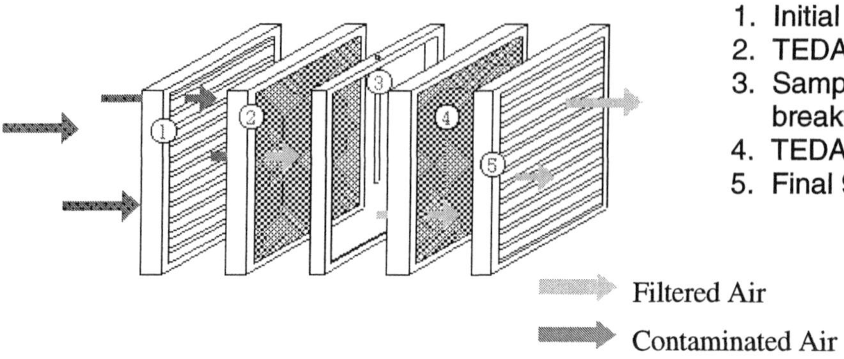

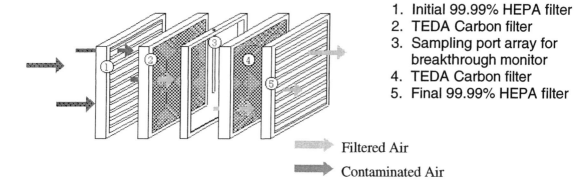

1. Initial 99.99% HEPA filter
2. TEDA Carbon filter
3. Sampling port array for breakthrough monitor
4. TEDA Carbon filter
5. Final 99.99% HEPA filter

Figure 2. Air Flow Pattern of High Containment Filtration System

Figure 3. Transportable Class III Glove box/Filtration System for NBC Material Assessment

Secondary Containment Facilities

Joint Service Installation Protection Project

Several strategies have been employed by the ECBC to address mobile and modular secondary containment facilities. In 2002, the Joint Program Executive Office for Chemical Biological Defense (JPEO-CBD) defined a requirement under the Joint Services Installation Protection Project (JSIPP) to field semi-mobile laboratory prototype systems to six military installations. The JSIPP was managed by the Defense Threat Reduction Agency (DTRA) and was a pilot project designed to enhance emergency response capabilities for chemical, biological, radiological, nuclear, and high explosive (CBRNE) events on military installations. Timelines associated with this project were extremely aggressive.

In early 2003, an agreement was established with the JPEO-CBD to design, fabricate, integrate, validate, and assist in the fielding of these six semi-permanent laboratory systems. The laboratory systems were integrated into 60-ft. box trailers and were designed to provide high throughput biological sampling and analysis with a Biosafety Level 2 capability. The six laboratory systems were in place and operational by the end of 2003, where they provide a critical function in the nation's biological counter-terrorism posture. They are deployed at our nation's most critical power projection installations. Terrorist or state parties eager to disrupt deployment of U.S. troops to high-tension areas around the world could target these installations. (See Figure 4).

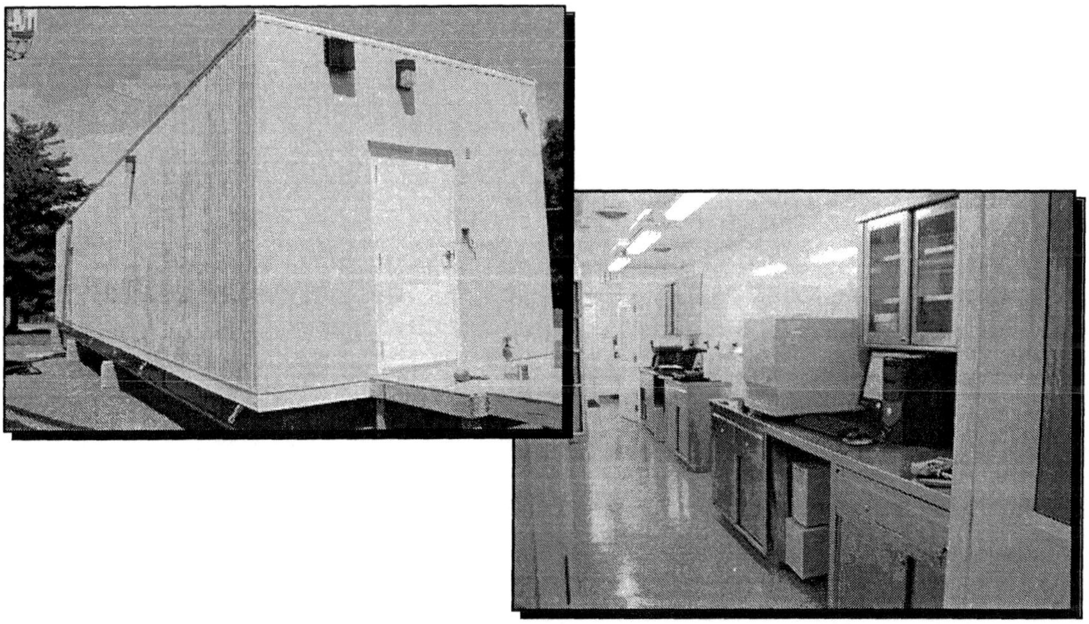

Figure 4. BSL 2 Biosafety Semi Transportable Laboratory

Follow the Air

From the beginning, the engineers on the mobile lab team agreed that there were many aspects of fixed laboratories that could not be extended to mobile facilities. We considered things such as weight distribution, power conservation, limited space for supplies, waste streams, fuel, potable water, generators, air filtration, redundant HVAC systems, and positioning and mounting all devices. Since contamination of a mobile laboratory will halt all work in the entire facility, avoidance of it is the highest priority. We therefore rely heavily on procedure, separation of analytical operations, where possible, and construction materials that are easily decontaminated.

Because we are mobile, we need to employ lighter weight construction materials such as plastics in lieu of slow curing concrete. We chose Fiberglass Reinforced Plastic (FRP) for the wall and ceiling material with as few seams as possible, essentially for its ease of decontamination. The floors are poured epoxy with raised coved splashguards to seal off and contain liquids throughout. It is essential that these materials have a good memory so they will not split or crack while in transport. The skeletal material is either aluminum or wood depending on the size of the platform. We found that the larger laboratories such as the JSIPP (60' x 12' x 14' wheeled transportable trailer) tend not to move as often and so weight of the facility does not play such a large role; they were constructed with wood studs. The more mobile laboratories are constructed with aluminum studs, which are more expensive but will endure years of traveling to different venues. Laboratory design specifications that are considered include:

- BSL 2
- Class 10,000 (ISO Class 7) "Clean room" conditions
- Easily decontaminated
- Interior seamless design
- Poured epoxy floor
- Environmentally controlled
- Pressure zoned for specific processes
- Stainless steel work surfaces
- Wide area network capability
- Two phone lines
- Steel structure 55lb/linear ft.
- Standard truck tires
- Coved corners
- Marine edges, coved corners
- Splash guards
- Double door interlocks in both entrances
- Video security cameras
- Intercoms (internal and external)
- UPS backup for all hoods and lab benches
- Interlocking doors
- Local area network backbone

Aside from the construction materials, the laboratory system is essentially designed around the air, its quality, and its movement within. All the doors and accesses are double gasketed. The entrances are configured with interlocking double doors to preserve air pressure within. The doors are electronically controlled so that no two doors can be opened at once. And, the interstitial rooms are maintained with a negative pressure and are well ventilated through HEPA filters. All the wall penetrations are caulked and covered. Every effort must be taken to contain the air within the filtration pattern. We make use of redundant HVAC systems that mount on the exterior of the laboratories. They are wired so that the blowers of both systems run continually and the cooling and heating are provided by only one system, alternating every seven days. All of the air is filtered through a large HEPA bank or HEPA/Carbon bank depending on the safety level of the laboratory and is usually sent to the opposite end of the laboratory where it is drawn back towards the return vents. We routinely achieve air quality far greater than class 10,000. A great deal of the air is exhausted through the safety cabinets and hoods where it is filtered and passed out, and a large portion of the remaining air is recirculated. We find this extremely efficient with the limited energy that we can provide for the laboratory. To this end, we find that single pass air is not required and does not compromise the safety of the laboratory staff within.

As a preplanned product improvement to the JSIPP laboratory systems, an environmental enclosure for a biological robotic system was designed and fabricated. This enclosed, laminar-flow biological safety cabinet was designed and delivered in less than 5 weeks after which it was installed, verified, and validated. The company that produces the robotic system, Beckman Coulter, is pursuing purchasing quantities of the enclosure from the fabricator and making it one of their suggested accessories with the robotic system.

Independent assessment/review of this system was established by the Department of the Army (SAR, 2003) and an independent third party (Richmond, 2004).

Combination of Class III Primary and Class II Secondary Containment into Mobile Facilities

Food and Drug Administration (FDA) Laboratory System

Through an Interagency Agreement signed in late 2002, the ECBC partnered with the Food and Drug Administration (FDA) to build one chemical and one microbiological mobile laboratory system. Each system is comprised of three platforms. One sample receipt/preparation (BSL2/BSL3), one analytical, and one RV-type vehicle for administrative and rest activities. By building and fielding mobile laboratories, the FDA will be capable of increasing their presence at ports of entry to our country. Secondly, and as importantly, these same mobile laboratories will be capable of responding to potential terrorist incidents in the contiguous United States.

As part of our policy and procedures, we attempt to mitigate contamination and cross-contamination by keeping sample materials under environmental controls where practical. Additionally, we physically separate the sample preparation from the analytical portion either by walls and differential pressure zones or by employing a completely separate mobile facility, as is the case with the FDA's biology and chemistry laboratories. Their mission is two-fold in that it necessitates that they have the ability to accept unknown, potentially dangerous materials as well as offer a continuing analytical surveillance of food and materials entering the United States ports-of-entry (POE).

The first facility in each of the FDA Laboratory Systems is the sample preparation portion of the laboratory. Here, the samples have two methods of ingress; either by way of the outside door into a small anteroom where a login procedure is performed or by an external passageway into an airlock connected to a Class III biological safety cabinet (Glove-box equipped with TEDA Carbon Filtration), also located in the anteroom. Standard screening methods are performed at this stage; dilutions are prepared and the sample can either be extracted back into the anteroom or placed into another airlock connected with a fume hood or Class II biological safety cabinet depending on the type of laboratory. The fume hood is located in a separate room, and all the doors of the airlocks have interlocks so that no two doors can be opened at once. The air entering through the glove-boxes in our preparation laboratories pass through a series of three HEPA filters prior to a bank of two thick beds of carbon with an air sampling system between them. A pre-filter follows them where the air is dispersed to the outside. The system is completed by a bag-in/bag-out capability.

The air moving within the laboratory travels against the direction of the sample with an estimated exchange rate of 1 air exchange every 1 to 2 minutes. The furniture is stainless steel throughout for ease of decontamination. The two rooms are pressurized negatively with the sample entry area being the most negative. This is continuously confirmed by pressure monitors placed on the walls. Aliquots of samples are processed in this lab and only enough material adequate for analysis is transferred to the analytical laboratory, thus lessening the chances for contamination. The analytical laboratories contain Class I and II cabinets for further processing of the samples if required.

Independent assessment/review of this system by the Department of the Army and an independent third party is pending. (See Figure 5).

Conclusion

Clearly, our approach to analytical analyses has evolved. We have gone from using our human senses (at the inception of mankind) to the highly sophisticated "Gold Standard" fixed-site laboratory. And now we are on our way back. Immediate and decisive results are more important than ever. Whether it is to protect our environment, our warfighters, our borders, our families, homeland security in general or space exploration specifically, the importance of the role of on-site analysis is evident and is our future.

The evolution of field analyses has led to the evolution of new engineering strategies that clearly follow prudent practices although they may be viewed as unusual. These strategies are addressing new and urgent requirements that provide customers with turnkey solutions to real-world requirements, when and where they need them. As you can see from this picture, we've come a long way from where we started and the future is limitless. (See Figure 6)

Figure 5. FDA Laboratory System

The Mobile Laboratory and Kits Team is a subordinate organization within the Edgewood Chemical Biological Forensic Analytical Center. The Edgewood Chemical Biological Forensic Analytical Center is an organization that operates within the ISO 9001:2000, ISO 17025, and ISO 14000 quality system.

Figure 6, Supporting the NATO 50th Anniversary
Washington,D.C.

References

Henry, C.E., Heyl, M., Reutter, D, Diez, H. Landy, K, (1997). "Super toxic analytical glove-box system (STAGS) (pp. 59-63). In M. Heyl & R. McGuire (Eds.) Analytical Chemistry Associated with the Destruction of Chemical Weapons.

Letter dated January 30, 2004 from Jonathan Richmond, PhD, Biosafety Consultant.

Novad, J.J., & Coale J.N. (1996). Edgewood RDE Center Support to the Alternative Technology Program, ERDEC-SP-053, Final Report Sep 94-Nov 96. Defense Technical Information Center AD-A331 136.

Richmond, J.Y., & McKinney, R.W. (1999). (Eds.). Biosafety in microbiological and biomedical laboratories 4th Ed. Washington, DC: U.S. Government Printing Office.

SAR (Safety Assessment Report) for the Joint Services Installation Pilot Project Semi-Mobile Prototype Laboratory System (JSIPP), Edgewood CB Center Risk Reduction Office, May 2003.

U.S. Patent No. 5,730,765: Super Toxic Analytical Glovebox System, Henry, Charles E.; Heyl, Monica J.; Reutter, Dennis J. (24 Mar 1998).

U.S. Patent No. 6,428,122: Portable Glove box & Filtration System, Henry, Charles E.; Heyl, Monica J.; Reutter, Dennis J. (6 Aug 2003).

Chapter 7

Management of Multihazardous Wastes from High Containment Laboratories

Edward H. Rau

Division of Environmental Protection, Office of Research Facilities Development and Operations, National Institutes of Health, U.S. Department of Health and Human Services, Bethesda, Maryland

Keywords:

Alkaline hydrolysis, animal carcasses, biocontainment laboratories, biosafety, decontamination, diagnostic and research laboratories, biosecurity, hazardous waste, incineration, infectious waste, medical wastes, mercury, mixed wastes, multihazardous waste, pollution prevention, prions, radioactive waste, and waste minimization.

Abstract

Small quantities of wastes contaminated with more than one type of hazardous material, referred to as mulithazardous wastes (MHW) are generated by biomedical research facilities. These wastes are often difficult to manage because of their sporadic patterns of generation and complex composition; the framework of myriad, often inconsistent and sometimes conflicting requirements of the several agencies regulating their management; limited treatment options and extreme disposal costs. Increased generation of these wastes from high containment facilities may be anticipated, and requirements for on-site inactivation and security of high-risk biohazardous agents in these facilities may exacerbate difficulties in management of other hazardous constituents in MHW by further limiting treatment and disposal options. Some mulithazardous waste streams such as contaminated animal carcasses, radioactive chemical wastes (mixed waste), mercury and prion contaminated materials, and wastewater from biological waste inactivation tanks are of particular concern. In this paper the major issues affecting MHW, the impacts of operational requirements for containment facility operations on their management, and potentially problematic waste streams will be reviewed. Some strategies, primarily developed from experience with MHW generated from biomedical research laboratories at the National Institutes of Health, which may be applied to management of MHW from high containment facility operations, will be presented. Implementation of these strategies should significantly reduce or eliminate management burdens and costs associated with MHW from containment facilities, improve security and assure protection of the environment.

Introduction

Multihazardous wastes (MHW) are wastes that exhibit any combination of chemical, radioactive or biological hazards (NRC 1995: NIH, 2004). Only biomedical facilities generate MHW that exhibits all three types of hazards. Low Level Mixed Waste (LLMW), generally referred to simply as mixed waste, is a specific subset of the universe of MHW that contains chemical constituents or exhibits characteristics as a composite that cause it to be regulated as hazardous waste under Resource Conservation and Recovery Act (RCRA), and radioactive materials regulated by the Atomic Energy Act. (Since the term "mixed waste" has a specific meaning defined by regulations it should not be used to refer to other MHW).

The need to develop measures to protect public health from bioterrorism threats and emerging diseases requires research using high-risk and high-consequence pathogens. Several major new high containment (biosafety level 3 and 4 (BSL-3 and BSL-4) laboratories will be constructed in the United States and existing facilities will be renovated to fill the need. Operation of these laboratories is likely to generate MHW with biohazardous components. Working with the high-risk pathogens handled in these facilities will require enhanced security, containment and on-site inactivation, adding significant complexity to management of MHW. During the same period that new containment laboratory capacity is being developed, more stringent environmental regulations, particularly for contaminants in wastewater discharges, are likely to be promulgated increasing compliance burdens.

In this paper MHW management issues and strategies for improved management will be reviewed. Much of the discussion presented here was drawn from experience at the intramural research laboratories of the National Institutes of Health (NIH). In excess of 10,000 containers of mulithazardous waste have been generated by these laboratories, characterized, treated and disposed of in the last ten years without incident. It should be understood that emphasis of research and associated procedures to be carried out in high containment laboratories may differ significantly from the majority of work currently performed in NIH laboratories, which is conducted in BSL-1 and. BSL-2. This may affect the characterization of wastes generated and selection of practices best suited for their management.

Multihazardous Management Issues

MHW typically comprises only a very small fraction of the total amount of waste generated by biomedical research laboratories yet it often consumes an inordinate percentage of management time and operating funds budgeted for waste disposal services, and presents a high potential for incurring regulatory violations and penalties. These difficulties primarily arise from a misfit of regulations, scant information on best management practices, and limited options for disposal.

Regulatory Framework

Management of MHW is complicated by an array of federal, state and local regulations that are often conflicting, unnecessarily burdensome, and inconsistent with the relative risk of each hazardous constituent in the waste. Regulation of MHW by multiple agencies greatly reduces the flexibility generators needed to carry out best practices for on-site treatment and disposal of these wastes, discourages development of off-site commercial disposal outlets, and contributes to prohibitively high disposal costs.

Problems with the current regulatory framework stem from several factors (NRC, 1995; NCRP, 2003; Rau, et al., 2000 ;). Most waste regulations were developed by each agency for the single hazard wastes under their respective regulatory authorities. The impacts of their requirements on management of constituents regulated by other agencies within the waste received little, if any consideration as regulations were promulgated. Secondly, most waste management regulations were developed to govern management of large, single hazard waste streams from industrial operations that tend to be generated repeatedly and in large volumes. Laboratory waste streams are typically small in volume, generated sporadically by highly variable processes, often unique, and of highly complex composition.

While there have been some recent improvements to the regulatory framework for LLMW, many of the problems with regulations governing MHW persist. Severe penalties can be assessed for noncompliance, particularly with regulations applicable to hazardous wastes and biohazardous wastes subject to select agent regulations...

Scarcity of Information

Scant information is available on minimization and management methods for most types of MHW. Better communication of the available information and new findings would facilitate improved waste management. Recommendations have been made to improve transfer of information on biomedical waste minimization and treatment technology among all stakeholders (Rau, et al., 2000) and to establish a national clearinghouse for this information (Barker, et al., 2000). Enhancement of existing regulations and new laboratory security regulations protecting site-specific information, identification of individuals who work with highly infectious agents and toxins, and the nature of agents present in facilities, may impede these efforts.

Few Disposal Options and Extreme Costs

The small volume of MHW generated by laboratories provides little incentive for waste disposal companies to make the large investments necessary to obtain multiple licenses and permits, research and develop handling systems, services, and treatment processes. As a result, few or no commercial facilities accept some of these wastes, and disposal fees for acceptable wastes may be extremely high. Disposal costs are affected by many factors but typical disposal costs compiled in 1999 indicated that common types of MHW generated by biomedical research laboratories were 10-100 times higher than similar single hazard waste streams (Rau, et al,, 1999). These were direct disposal costs; they did not include additional costs that may be incurred for waste analysis and characterization required to obtain acceptance of waste at commercial disposal facilities; dedicated shipments; and other specialized services required to manage MHW prior to final disposal.

Minimization of MHW

The difficulties in managing MHW and extremely high disposal costs for some waste streams dictate the need to avoid generation of MHW and minimize the amount and toxicity of unavoidable wastes. Regulatory requirements also apply. The Pollution Prevention Act of 1990, the Resource Conservation and Recovery Act (RCRA)[2] and hazardous waste regulations[3] require all generators to minimize wastes and toxic releases to the environment. Recent Presidential Executive Orders applicable to federal agencies require them to implement pollution prevention and waste minimization across all mission activities including those performed by contractors (Clinton, 1998, 2000).

Wastes from research activities are often unique and no overall waste minimization methodology can be applied to the myriad waste generating processes. Some general strategies of broad application are listed in order of preference from the standpoint of environmental protection:

1. **Source reduction or waste avoidance.** Where feasible, replace hazardous chemicals with less hazardous compounds, use alternatives to radioactive methods, and lower biohazard risk organisms to reduce or avoid generation of MHW.

2. **Segregation.** Proper segregation of wastes can greatly reduce generation of MHW. Chemical, radioactive and biohazardous wastes should never be commingled. All radioactive wastes regardless of form should be segregated by radionuclide to the maximum extent possible. If multiple label experiments are performed, the waste should be segregated by half-life to the greatest extent possible. Wastes containing long-lived

---

[2] Sec.1003[b], 1984

[3] 40 CFR 262.41(a)(8)70

radionuclides should never be mixed with short-lived wastes that can be held for decay in storage.

3. **Recycle wastes where feasible.** Few MHW waste streams are recyclable directly without processing. Liquid scintillation fluids can be used as fuel in heat recovery facilities. In some cases, materials such as plastics can be recovered from MHW that has been treated to remove or inactivate hazardous constituents. Alkaline hydrolysis digester waste may have potential applications as fertilizer or as a nutrient supplement in biological wastewater treatment systems.

4. **Treat unavoidable wastes to reduce volume, toxicity and biohazards.**

5. **Dispose of treated wastes in a manner that is most protective of the environment.**

Recent publications provide compilations of information on waste minimization strategies for laboratories (NCRP, 2003; NCCLS, 2002; USEPA, 2000).

Treatment and Disposal of MHW

On-Site Treatment

Some on-site treatment of MHW is usually required to reduce multiple hazards to a single hazard such that the resulting single hazard waste can be shipped to off-site recycling, treatment, storage, and disposal facilities. Treatment and disposal plans for MHW waste streams are usually made on a case-by-case basis and often require the expertise of a multidisciplinary team including biosafety professionals, chemists, and health physicists. Plans consist of a specific sequence of steps based on the constituents in the waste; waste form, quantity and generation frequency; availability of equipment and validated treatment methods; acceptance requirements of off-site facilities; and regulatory requirements applicable to each type of hazardous material in the waste.

In some cases more than one type of hazardous constituent in MHW can be treated at the same time:

- Sodium hypochlorite used for disinfection can also inactivate toxins and can detoxify some toxic organic chemicals;

- Ultraviolet peroxidation used to degrade ignitable and toxic organic chemicals in aqueous wastes is probably capable of inactivating infectious biohazardous agents; and

- Incineration destroys organic chemicals and infectious agents, and can reduce the volume of radioactive wastes, potentially converting a chemical-radioactive-biological waste to a simple radioactive waste.

Biohazardous Agents and Other Regulated Biological Materials

Biohazardous and regulated characteristics of wastes relating to biological materials include the presence of viable infectious agents, toxins, recombinant DNA, nucleotides, and other biological constituents regulated as select agents; potential for putrefaction, odor generation, pest or vector attraction; and aesthetic or security-related characteristics of carcasses, anatomical parts, tissues, body fluids, and specimen identification information that must be rendered unrecognizable to meet aesthetic, security or medical waste regulatory requirements before final disposal. Recombinant DNA, components of biological expression systems (host cells and viral vectors potentially contaminated with replication-competent viruses), transgenic organisms, and other discarded materials from recombinant DNA technology may also be biohazardous constituents of multihazardous waste.

On-site treatment usually focuses on inactivation of infectious agents. Treatment of other biohazardous and regulated properties related to biological materials, and final disposal is usually accomplished off-site by incineration or other processes appropriate for the nonbiological hazardous materials remaining in the waste.

Strategies employed for management of MHW may require some modification for applications in high containment facilities:

- Requirements for inactivation of biohazardous agents in wastes before exiting containment (Richmond, McKinney 1996; USAGFD, 2001; WHO, 2004) may limit flexibility in determining the most appropriate sequence of treatment steps;
- Potential incompatibility with inactivation systems must be considered, e.g., autoclaves not appropriate for processing MHW containing volatile radionuclides or chemicals; and
- There may be difficulties in operating and maintaining treatment systems, and providing repair personnel access for work within containment zones.

Hazardous Chemical Constituents

The presence of specific hazardous chemical compounds in MHW may result in it being regulated as a hazardous waste or LLMW under USEPA regulations if the constituents appear on any of four lists, or the waste as a composite material exhibits certain defined hazardous characteristics of ignitability (corrosively, reactivity or leachate toxicity)[4]. Wastes that are not regulated as hazardous may still be subject to EPA, state and local regulations pertaining to discharges into wastewater systems, or other environmental media.

---

[4] 40 CFR 261

The primary goal of treating chemical constituents in mulithazardous waste is to eliminate characteristics that result in the waste being regulated as a hazardous waste such that it can be accepted by off-site treatment and disposal facilities that are not EPA-permitted for hazardous waste for additional treatment of other hazardous constituents or final disposal. On-site treatment can also assist generators in complying with regulatory requirements to minimize the volume and toxicity of hazardous wastes.

There is a widely held misconception that regulated hazardous wastes cannot be treated in laboratories without EPA permits. There are several exceptions to the ban on treatment without a permit which may be used to improve management of MHW.

Most MHW generated by laboratories is generated sporadically and in small volumes and thus may be amendable to treatment in containers or small accumulation tanks. Hazardous wastes may be treated in containers or tanks without a permit if in conformance with minimum management standards[5]. Container management requirements are summarized below:

1. Containers must be made of or lined with materials that will not react with, and are otherwise compatible with the hazardous wastes to be treated or stored.

2. Areas where containers are stored are inspected at least weekly, looking for leaks and for deterioration caused by corrosion or other factors.

3. Containers are kept closed except when necessary to add or remove waste.

4. Wastes in containers that are leaking or not in good condition are transferred to containers in good condition.

5. Treatment and storage time does not exceed the applicable time limits (usually 90 and 180 days if the waste has been removed from the satellite accumulation area..

6. The laboratory has a waste analysis plan if they are treating their waste to meet land disposal requirements.

Examples of treatment procedures that can be carried out in accumulation containers include precipitating toxic heavy metals from solutions and oxidation-reduction reactions. Biological treatment of hazardous organic chemical constituents in mixed wastes from laboratories has been successfully demonstrated (Wolfram, et al., 1997).

---

[5] 40 CFR 262.32 and Subparts I and J of Part 265.

There are three other options for treatment of chemical content that do not require a permit:

1. Wastes that are regulated as hazardous only because they are corrosive may be neutralized and managed as nonregulated waste[6].
2. Treatment that occurs as part of the process - adding an extra step to a procedure to render the byproducts nonhazardous.
3. Treatment that is part of a recycling process. Generally, laboratory waste streams that are generated in larger quantities provide the greatest opportunity for recycling. For laboratories, spent solvents are often among the highest volume hazardous waste streams generated. Recovery of solvents from histology procedures and silver from photoprocessing wastes are common examples of recycling processes practiced in laboratories that are not subject to permitting requirements.

Many toxic chemicals used in biomedical laboratories are not regulated as hazardous waste because they have not been specifically listed in regulations defining hazardous waste and they to not exhibit any of the characteristics of unlisted hazardous waste[7]. All wastes should still be treated and disposed of safely and in a manner that is protective of the environment. However, these nonregulated chemical wastes can be managed with greater flexibility and are not subject to storage time limits. This allows for accumulation in storage areas outside the laboratory's satellite accumulation areas without regard to time limits.

Few facilities have developed fully permitted and licensed systems for treating hazardous chemicals in MHW because of the extreme regulatory hurdles that must be passed to obtain USEPA permits. Most laboratory facilities do not generate enough treatable waste to justify the effort.

A modified version of a commercially available ultraviolet peroxidation treatment system has been developed by the NIH and used to degrade ignitable and toxic organic compounds in high-volume aqueous mixed waste (Rau, 1997). The system uses hydrogen peroxide in the presence of a catalyst and ultraviolet light to oxidize organic compounds to carbon dioxide and water. The demonstrated removal efficiency for targeted volatile and semivolatile compounds is in excess of 99.99%. No hazardous air emissions or residues are produced by the treatment process. Treated wastewater from the UVP system can be discharged

---

[6] 40 CFR 264.1(g)(6) and 270.1(C)(2)(v))

[7] 40 CFR 261

to the sanitary sewer. Concentrations of toxic organic compounds in a typical batch of waste before and after UVP treatment are shown in Table 1.

Table 1. Reduction of Toxic Organic Compounds in a Batch of Aqueous Mixed Waste Treated by Ultraviolet Peroxidation (UVP)

Target Contaminant	**Concentration (mg/l)**	
	Pretreatment	**Post treatment**
bis-(2-ethylhexyl) phthalate	>18.0	0.000
Chloroform	614.3	0.0
Di-n-butylphthalate	0.000	0.0112
Dichloromethane	13.6	0.0
Ethylbenzene	7.1	0.0
Phenol	>1,380	0.0
Toluene	>1,480	0.0
Total Toxic Organics	>3,513	0.063

One of the major problems in managing MHW that is regulated as hazardous waste is dealing with storage time limits, which may not allow sufficient time to find disposal outlets or accumulate quantities of waste that may be disposed of in a cost-effective manner. Hazardous waste regulations require that wastes in central storage areas must be disposed of within set time frames. Most biomedical laboratory facilities are classified as large generators and as such are only allowed to hold wastes in storage areas for a maximum of 90 days. Major regulatory sanctions may be imposed on generators that exceed the storage time limit.

A solution to this problem may be to hold wastes at the point of generation in the laboratory where wastes initially accumulate and under the control of the operator of the generation process until disposal can be arranged. Such locations are commonly known as "satellite accumulation areas." As much as 55 gallons of hazardous waste or one quart of acutely hazardous waste listed in 40 CFR

261.33(e) may be held in containers at or near any satellite accumulation area for an indefinite time period and without permits[8].

Further information on minimization of chemical wastes from laboratories from is available in the online guide *Less is better* published by the American Chemical Society (ACS, 2004).

Radioactive Materials

The increasing use of alternatives to radioactive materials such as chemiluminescent dyes in essay procedures has greatly reduced procurement of radioactive materials and generation of radioactive wastes (Austin et al., 2002, NIH, 1995). However, radioactive materials will still be used in biomedical research for the foreseeable future. Laboratories typically use radioactive materials in the following applications (USEPA, 2000):

- Radioisotopes, usually in liquid form, are widely used as labels in assay procedures. Commonly used isotopes are ^{3}H (tritium), ^{14}C, ^{32}P, ^{33}P, ^{35}S and ^{125}I.
- Contrasting agents used in electronmicroscopy may contain uranyl acetate, uranyl nitrate, and thorium nitrate which are radioactive.
- Sealed radioactive sources are used in measuring devices. Examples are ^{63}Ni used in gas chromatographs, and ^{210}Po in static eliminators. Since these are in sealed sources, they are not likely to be present in wastes.

Research facilities engaged in imaging studies may use additional shorter lived radionuclides such as ^{99}Tc, ^{111}In, ^{135}I. Accelerator produced radionuclides such as ^{201}Tl, ^{123}I, and ^{18}F may also be used; disposal of these is not subject to USNRC disposal regulation but may be regulated by the states.

Options for treating radioactive materials in MHW are largely limited to holding short-lived radionuclides for decay in storage. With the exceptions of tritium, ^{14}C and the contrasting agents, the isotopes likely to be found in wastes have short enough half lives to be held for decay in storage and then managed without regard to their radioactivity. This allows management of the decayed waste as nonradioactive waste and obviates the need to ship the waste to costly off-site disposal facilities. Freeze-drying, compactors, and incineration can also be used to reduce the volume of waste shipped off-site.

It should be noted that some short-lived radionuclides which may be used in biomedical research facilities such as ^{111}I often contain small amounts of long-lived radionuclidic impurities which may complicate simple waste management procedures, and disqualify wastes containing them from decay-in-storage (DIS) programs. For these radionuclides, gamma spectrometry can be used to aid

---

[8] 40 C.F.R. 262. 34(c)(1)

waste segregation and final management decisions on low level radioactive waste management (Salako & DeNardo, 1997).

Problematic MHW Waste Streams

Animal Carcasses

The disposal of carcasses is a problem faced by many animal facilities (Richmond et al., 2003). Traditionally, they have been disposed of by incineration either locally or off-site by medical waste disposal companies. In recent years this practice has become subject to increased regulatory restrictions and continued access to incineration is uncertain. Many commercial disposal facilities have closed incinerators and are now using alternate technologies for treating solid medical wastes such as grinding and microwave heating or chemical disinfection. Most of these alternate technologies are not suitable for processing carcasses.

In most cases, animal carcasses from research facilities are minimally subject to regulation as medical waste and may contain infectious agents or toxins. However, carcasses from animals contaminated with radioactive materials, regulated levels of certain toxic chemicals or fixatives such as formalin may be considered MHW and subject to multiple regulatory requirements. Carcasses regulated as mixed waste are rarely encountered. However, many facilities generate radioactive carcasses and management of these carcasses has become increasingly problematic.

Generally, facilities manage carcasses with short-lived radionuclides by holding them for decay-in-storage and then processing as nonradioactive medical waste. This requires facilities to have on-site holding capacity, usually in freezers for storage of carcasses during decay. Recently, the Nuclear Regulatory Commission eliminated a policy requiring short-lived radioactive wastes to be held for ten half lives before they could be monitored and released as nonradioactive waste (NRC, 2004). The new policy is more risk-informed and performance based, allowing wastes to be released as long as a final survey determines that radiation levels in the waste cannot be distinguished from the background levels. This may reduce the need for carcass storage capacity and expedite disposal.

Carcasses with long-lived radionuclides cannot be decayed in storage and are generally shipped off-site to radioactive waste incineration or landfill facilities. (There is an exemption for carcasses with below the 1.85 kBq (0.05 µCi) of tritium or ^{14}C per gram and these may be managed as nonradioactive waste)[9].

---

[9] 10 CFR 20.306

Currently, few commercial facilities accept radioactive carcasses and disposal costs are very high.

Management of MHW carcasses generated within high containment areas is further complicated by the need to inactivate infectious agents before the waste exits the containment area. Solid wastes such as carcasses are usually inactivated and passed out of the area through autoclaves and then treated by a terminal process such as incineration. Autoclaving has several limitations for MHW carcasses:

1. The efficacy of autoclaving carcasses as a single step inactivation process for infectious agents in larger animal carcasses has not been demonstrated.

2. Long processing times are required to assure attainment of core temperatures necessary to inactivate pathogens.

3. Autoclave treatment may not be appropriate for carcasses containing volatile radionuclides, chemicals or fixatives.

4. Autoclaving does not render carcasses unrecognizable nor does it inactivate some toxins. Therefore, autoclaved carcasses may still be regulated as medical waste, be subject to biosecurity regulations, and will be unacceptable at radioactive and hazardous waste disposal facilities.

Hot alkaline hydrolysis carcass digestion systems originally developed for disposal of radioactive carcasses (Kaye & Weber, 1992) are now commercially available and may offer a potentially advantageous alternative to autoclaving and incineration of MHW carcasses:

1. Radionuclides are solubilized or converted to biologically-disbursable forms, which meet requirements for discharge to wastewater systems (provided activity concentrations are below NRC discharge limits).

2. The efficiency of the process in inactivating a wide variety of infectious agents has been demonstrated. Carcasses from several species were processed with cultures of representative pathogens including Staphylococcus aureus, Mycobacterium fortuitum, Candida albicans, Bacillus subtilis, Pseudomonas aeruginosa, Aspergillus fumigatus, Mycobacterium bovis BCG, MS-2 bacteriophage, and Giardia muris and were processed in a digestion unit. No evidence of infectivity was found in the digestion products (Kaye et al., 1996).

1. Reaction conditions in alkaline hydrolysis destroy common tissue preservatives such as formalin, glutaraldehyde, and phenol, and should be capable of inactivating biological toxins, nucleic acids and nucleotides, antineoplastic agents regulated by EPA as hazardous waste, nucleic acids, and antineoplastic agents (Kaye & Weber, 2004).

2. Digestion renders carcasses unrecognizable (except for fragile bone residues that can be disposed of as nonregulated waste).

3. Air emissions associated with incineration are eliminated.

4. In most cases, digestates from the process can be discharged directly to sanitary sewerage systems, obviating the need to ship wastes off-site.

Use of alkaline hydrolysis systems has some potential disadvantages:

1. Handling of caustics such as sodium hydroxide and use of these materials in hot pressurized systems presents physical and chemical hazards.

2. Digesters have not been operated in high containment; it is uncertain how such units will be installed, operated and maintained in these areas.

3. The biological oxygen demand and nutrient concentrations (nitrogen and phosphorus) in digestates may exceed local discharge limits and require further processing, drying or off-site disposal.

Few methods have been reported to reduce the concentrations of long-lived radionuclides in carcasses. Under specific circumstances concentrations in animal carcasses can be reduced by an average of 88% by use of the exsanguination method of animal euthanasia. Animals undergoing exsanguination are fully anesthetized, and the blood is removed resulting in hypovolemia. In situations where radioactive materials are used as part of a research protocol that remain predominantly suspended in the blood, the exsanguination procedure can result in a significant lowering of residual radioactivity content. This reduction can greatly affect the types of waste management and minimization options that can be subsequently applied.

For animals in protocols with tritium or ^{14}C such reductions can be a very significant advantage because they may reduce the activity concentration in the carcass below the USNRC's exemption limit, allowing disposal of the carcasses as nonradioactive waste. Although the exsanguination procedure can result in significant waste minimization opportunities in certain circumstances, this should not be the rationale for its use. Rather, the method of euthanasia should be based exclusively on sound animal care and use principles, and waste management strategies should then be made following that decision (Costello et al., 2000).

Low-Level Mixed Wastes (LLMW)

LLMW are the most expensive waste streams generated by biomedical research facilities to manage and dispose of. Some LLMW streams have few or no current disposal options. As a consequence, some laboratories have placed restrictions on the generation of mixed waste, requiring investigators to develop alternative research methods and procedures that may, in some cases not be as effective as isotopic methods.

Experience at NIH illustrates the potential impacts of LLMW on biomedical laboratory operations. In the mid-1990s, the volume of LLMW generated by NIH facilities increased ten fold due to growth of the intramural research program, increased usage of molecular biology techniques that generate mixed waste, and changes in regulatory definition of hazardous waste that required more low level radioactive wastes to be managed as mixed wastes[10]. During the same period options for off-site disposal of many of the mixed waste streams were very limited or extremely costly - in the range of $15,000 to $50,000 per 55-gallon drum. Much of the waste had to be stored until on-site treatment methods could be developed and permitted. Following implementation of a comprehensive minimization program generation of these wastes was reduced by over 99%. A detailed review of the NIH mixed waste minimization and management program is available (Rau, 1997).

Key to preventing LLMW generation is training, ensuring that all investigators authorized to use radioactive materials are aware of applicable regulatory requirements, the difficulties and costs associated with managing LLMW, and waste avoidance methods (Austin et al., 2002, Holcomb, et al., 1993; Holcomb, 1995). A video film on minimization and management of mixed waste was prepared by the NIH for use in training (NIH, 1992).

Generation of large volumes of LLMW from high containment facilities is not anticipated:

- Most molecular biology techniques that generate LLMW are likely to be performed in BSL-1 and BSL-2 laboratories or at off-site facilities;
- There is increasing availability of alternatives to radioactive materials for most common essays and research procedures; and
- Activities within containment areas primarily involve work with animals, accelerator produced radionuclides and microbiology procedures that do not generate LLMW.

While the amount of LLMW generated by these facilities is likely to be minimal, some generation will occur. Even generation of very small quantities e.g., 100ml per year that cannot be disposed of may result in the imposition of serious regulatory sanctions. Meeting the legal and regulatory requirements to generate, use, store, treat, and dispose of mixed waste can be costly and difficult. The regulations that govern mixed waste, from creation to final disposition, are

[10] Most important change was the adoption of a concentration limit of 6 mg/l for chloroform. Liquid radioactive wastes exceeding this threshold had to be managed as LLMW. Low levels of chloroform are ubiquitous in biomedical wastes. It is used directly in nucleic acid and lipid extraction procedures; a product of the incomplete degradation of trichloroacetic acid, which is widely used as protein precipitant; and a reaction product of sodium hypochlorite and organic materials.

extensive and complex to implement and have created unnecessary burdens on biomedical research (NIH, 1999; NCRP, 2003; Rau, et al., 2000). Noncompliance with the multitude regulatory requirements can lead to serious repercussions including harm to individuals and the environment, fines and penalties against a facility, and/or suspension of a facility's operations (Hageman, 2002).

Many of the problems confronting research facilities in managing LLMW result from the dual regulatory framework of sometimes conflicting hazardous wastes and radioactive waste regulations. The Environmental Protection Agency (USEPA) enforces hazardous waste regulations. The Nuclear Regulatory Commission (USNRC) is the primary enforcer of regulations governing radioactive materials in waste; the USEPA has a limited role. The USEPA and USNRC may authorize states to enforce their respective regulations, and states may adopt additional regulations that are more inclusive or stringent than those of the USEPA and USNRC. State hazardous waste regulations are enforceable on federal installations; state radioactive materials regulations are not.

Problems with applying the regulatory framework for LLMW to wastes generated by laboratories have been evident for a long time. In 1981, the USNRC relieved some of the laboratory mixed waste problem by allowing liquid scintillation fluid (LSF) with less than 1.85 kBq/g of ^{3}H or ^{14}C to be disposed of without regard to radioactivity. Thus ignitable LSF below this limit need not be managed as a mixed waste but only as a hazardous chemical waste (NRC 1995). Recent trends in generation of mixed waste from biomedical laboratories have significantly diminished the value of this exemption. The majority of mixed wastes now generated are from molecular biology procedures such as gel electrophorsis, not liquid scintillation counting and therefore are not exempt (USDOE 1994,1995; Rau, 1997).

U.S. NRC policy did not establish a *de minimis* level for other radionuclides and types of laboratory radioactive waste. However, licensees can propose a license-specific *de minimis* level, below which mixed waste can be released for management as a chemical waste (NRC, 1996).

New regulations (USEPA 2001a, 2001b) provided increased flexibility to facilities for managing low-level mixed waste and naturally-occurring and/or accelerator-produced radioactive material (NARM) containing hazardous waste. LLMW was exempted from hazardous waste storage and treatment requirements as long as the waste is generated under a single NRC license, meets the conditions specified, and is stored and treated in a tank or container. In addition, LLMW that meets applicable treatment standards may be conditionally exempt from RCRA transportation and disposal requirements. This waste may be disposed of at low level radioactive waste disposal facilities that are licensed by NRC. The new rule also provides additional flexibility for manifesting these wastes when they are destined for disposal at such facilities. Although mixed waste meeting the applicable conditions is exempt from certain hazardous waste requirements, it must still be managed as radioactive waste according to NRC regulations. A checklist for compliance with the new regulation is available (Yoder, 2002).

Recently, the USEPA proposed another rule (USEPA, 2003) that could provide further relief from regulatory burdens associated with wastes containing extremely low levels of radioactivity. It is intended to:

- Improve the scientific characterization of wastes with low concentrations of radioactivity ("low activity wastes");
- Promote a more coherent and consistent framework for the disposal of these wastes, especially low-activity, mixed wastes (LAMW);
- Increase options for management of LAMW including allowing burial of hazardous waste in landfills not licensed to accept radioactive waste; and
- Provide the flexibility to allow laboratories and other waste generators to account for specific state or local regulatory constraints and economic considerations in determining whether they would choose to implement this disposal option.

Mercury-Containing Wastes

Mercury is probably the most potentially problematic contaminant in biomedical research facilities and waste streams generated by research activities. Elimination of mercury devices such as thermometers and sphygmomanometers, and mercury-containing reagents has been a priority of pollution prevention campaigns in many hospitals and health care facilities (H2E, 2004; Sustainable Hospitals, 2004), at the National Institutes of Health (NIH, 1997, 2004; Rau, 2003) and in other biomedical research organizations for the last decade. Several factors have driven these efforts. Mercury presents serious potential health hazards to employees, patients and laboratory animals, and is persistent, toxic, and highly bioaccumulative in the environment.

OSHA regulations[11] applicable to occupational exposures of employees, establish a Permissible Exposure Limit (PEL) for inorganic mercury in air of 1 $mg/10m^3$ (0.1 mg/m^3) or 1,000,000 ng/m^3 as a time weighted averaged over 8 hours (OSHA,1996). The Agency for Toxic Substances and Disease Registry (ATSDR) has developed a much lower chronic Minimal Risk Level (MRL) of 0.2 $\mu g/m^3$ or (200 ng/m^3) (ATSDR, 1999). Using different assumptions and risk factors, EPA has set an approximately equivalent Reference Concentration (RfC) for mercury at 0.3 $\mu g/m^3$ or 300 ng/m^3 (USEPA, 1988). A Suggested Action Level (SAL) of <1.0 $\mu g/m^3$ or $<1,000$ ng/m^3 extrapolated from the ATSDR and EPA values has been established which is considered acceptable for occupancy of areas after spills (ODH, 2004). The MRL and RFC are estimates of the maximum daily exposure level that is likely to be without appreciable risk of adverse, non-cancer health effects. The MRC, RFC, and SAL levels illustrate the extremely low levels of airborne mercury contamination that may be of concern. They are probably more appropriate than the OSHA PEL to apply in areas occupied by

[11] 29 CFR 1910.1000

patients and laboratory animals that may be more sensitive to its toxic effects and reside in affected areas for long periods of time.

Mercury spills of even a few drops in an enclosed indoor space can raise air concentrations to levels that may be harmful to health. Spills of larger volumes such as those used in sphygmomanometers can contaminate very large areas and require extremely expensive clean-up efforts. Releases in to the environment of only one pound of elemental mercury (about 34 mL) or one pound of waste containing as little as 0.2 mg of mercury per liter require immediate reporting to the National Response Center, state, and local authorities.[12] Mixed wastes contaminated with mercury, particularly as organometallic compounds, may have no treatment or disposal options (Rau, 1997). Mercury contamination is often widespread in older biomedical laboratory facilities and may account for the majority of decommissioning costs (Rau & Carscadden, 2004).

The availability of alternatives to mercury in virtually all health care applications and laboratory reagents has allowed replacement of mercury with minimal impacts on mission activities. Pollution prevention programs have been highly effective in reducing the use of mercury devices in facilities and the incidence of spills and the associated clean-up costs have been greatly reduced. However, even in these facilities, mercury contamination will continue to present operators with serious compliance challenges because of the emergence of extremely stringent discharge limits for mercury in wastewater and targeting of biomedical facilities and laboratories for enforcement actions when these limits are exceeded. The situation in Massachusetts illustrates the impact of these limits on regulated biomedical facilities.

In Massachusetts the Water Resources Authority (MWRA) imposed an effective sewer discharge limitation for mercury of 1.0 part per billion (ppb) from its regulated sources. Meeting the MWRA's standard for sewer discharge presented a formidable challenge for hospital laboratories because key substances used in research and diagnostic procedures, reagents in particular, often contained trace amounts of mercury that are usually not listed in the content descriptions. A work group was established in 1994 to examine the problem and develop strategies to determine the sources of mercury in wastewater and reduce the amount of mercury being discharged. Hospital participation in this process was coordinated through MASCO (a not-for-profit provider of services and technical assistance to Longwood Medical and Academic Area institutions), and involved the active participation of 28 hospitals. The results of extensive studies by the work group and MASCO have been reported (MASCO, 2004a) and are summarized below:

---

[12] 40 CFR 302.6, 355.40

1. Some of the more obvious sources of mercury, such as thermometers, manometers, and laboratory chemicals were readily identified by their MSDS to contain mercury.
2. Over 5,500 products used by hospitals and institutions were sampled and analyzed for total mercury content. Of those tested, approximately 781 were confirmed to contain some level of mercury. Seventy-two product analysis records indicated a mercury content of between 1 - 10 ppb and 469 records noting a concentration above 10 ppb.
3. Some of the higher concentrations were found in widely used commercial chemicals such as bleaches, soaps, cleaners, x-ray film processing chemicals, and wastewater pretreatment chemicals that are ultimately disposed in high volumes via sewage systems. Some laboratory reagents such as hematoxylins and acid zinc formalin, and products such as saline solutions that contained Thimerosal also had significant levels of mercury.
4. Testing procedures varied widely, depending upon the type of testing or research being conducted, making standardization of procedures that may use mercury-containing reagents exceedingly difficult.
5. Embedded tissues that had been fixed in mercury containing fixatives were found to leach mercury that contaminated other areas of the histology laboratories.
6. Trace amounts of mercury in wastewater tend to collect in the organic material (biomass) that may be present in waste piping systems and, as a consequence, can slough off into the wastewater stream at any time resulting in excursions above the allowed discharge limitation.
7. Microbial metabolism of mercury in plumbing biomass converts some of the mercury to highly toxic methylmercury, which can then be absorbed or bioaccumulated in the biomass or released in vapor form into the workplace atmosphere.

Biomedical laboratories should respond proactively by establishing a rigorous, source reduction program with these key elements:

1. Facility design specifications should strictly prohibit installation of mercury devices in building systems. Mercury free alternatives are available for most common applications such as thermostats, flow meters, manometers, switches, and high intensity discharge lights.
2. All laboratory chemicals, disinfectants, reagents, biologicals, animal feed and bedding, and other supplies should be screened for mercury content by sampling and analysis or checking existing product databases. MASCO provides a searchable computerized database listing approximately 8,000 chemicals used by hospitals and institutions (MASCO, 2004b). For about 800 listed products, the database includes the results of analytical testing

for mercury content. Procurement protocols should be established to require selection of products with the lowest available mercury content.

3. Only mercury-free tissue fixatives should be used.

4. Specifications for fluorescent tubes, which have some unavoidable mercury content should specify lighting tubes with the lowest available mercury levels and require the tubes to be enclosed in protective plastic sheaths to reduce the potential for breakage during installation, use, and removal.

The impacts of mercury pollution on public health are increasingly evident. In a recent survey it was found that approximately six percent of childbearing-aged women had levels at or above a reference dose, an estimated level assumed to be without appreciable harm (CDC, 2004). Attempts to reduce human exposure will undoubtedly result in even more stringent discharge standards. This is already happening in some areas. In EPA Region 5, a Water Quality Based Effluent Limit (WQBEL) 1.3 ng/l (parts per trillion) now applies to discharges of treated wastewater from Publicly Operated Treatment Facilities (POTWs). Many POTWs in the Region do not currently meet this standard and will have to develop and submit Pollutant Minimization Plans (PMPs) to come into compliance. EPA's recently issued guidance for development of mercury pollutant minimization programs by POTWs (EPA, 2004) recommends inclusion of individual facilities and targeted groups in PMPs. The guidance specifically lists medical facilities and laboratories as direct contributors of mercury pollution. Existing laboratories are likely to be targeted and in some states, such as Minnesota (MCPA, 2004) new dischargers will be required to comply with the WQBEL immediately. Designers of new biocontainment facilities should be aware of this potential requirement.

Aside from the environmental health and regulatory issues, mercury contamination in infectious disease laboratories may also be of concern as an extraneous variable that could potentially affect the outcome of research. Mercury can be a potent immunostimulant and suppressant, depending on exposure dose and individual susceptibility, and may increase susceptibility to parasitic, bacterial, and viral infections (Sweet, Zelikoff, 2001; Tchounwou, et al., 2003; Shen & König, 2001 et al.). For example, subtoxic exposure to mercury has been shown to enhance susceptibility to murine leishmaniasis (Bagenstose, 2001) and impair the murine response to Plasmodium yoelii infection and immunization (Silbergeld et al., 2000).

Avoiding all unnecessary uses of mercury and mercury contamination is even more critical in high containment facilities and any necessary uses warrant extreme precautions:

- Mercury spill response and decontamination requires specialized expertise that is not likely to be held by biomedical investigators trained and authorized to work in high containment areas.

- Monitoring of affected areas to determine the extent of contamination and the effectiveness of decontamination efforts requires the use of very expensive, sensitive equipment — portable atomic absorption spectrophotometers, which would be very difficult to bring into the area and decontaminate when exiting.

- Mercury-contaminated spill clean-up materials would be difficult to move out of the containment area because the usual thermal methods of inactivation before pass through such as autoclaving depend on thermal methods and would volatilize the mercury and cannot be used.

- Virtually all effluent from washing of surfaces and other wastewater generated within containment zones drains to cook tanks for biological decontamination. Any mercury contamination in these zones is likely to eventually enter the tanks. Given the extremely low levels of mercury allowed in releases, even very small amounts of contamination such as might occur from breakage of a single fluorescent light tube could raise the concentration of mercury in the tanks above allowable discharge limits.

- Treatment of wastewater to remove low levels of mercury contamination is probably not feasible.

Technologies presently available for use in removing low levels of mercury (e.g., a maximum of 1,000 ppb) to absolute minimum levels (less than 1 ppb) have been evaluated (MASCO, 2004). These included simple filtration, reverse osmosis, chemical precipitation/redox reactions, disinfection, membrane microfiltration, ion exchange, adsorption, and evaporation. They found that:

1. None of the technologies was individually or collectively capable of reducing the concentration of mercury in a facility's discharge to below 1.0 part per billion on a consistent or sustainable basis.

2. Some of the technologies have demonstrated abilities in removing 99.7% of the total mercury from the waste stream prior to discharge but the treated effluent still has mercury content at the 3 to 5 ppb level.

3. Initial pretreatment is required before these advanced techniques can be applied; all of which requires a significant amount of space and money to be installed.

4. There are many characteristics of biomedical wastewater streams that, if not controlled, can significantly and adversely impact some of the technologies that were investigated. For example, chlorine bleach, used as a disinfectant, can cause a rapid deterioration of the membranes used in nanofiltration and reverse osmosis based systems. Oil and grease can cause an almost immediate failure of ion exchange media. Organic materials and biological activity present in the raw wastewater may result in degradation of the activated carbon and premature failure of the media.

Radioactive Biohazardous Solid Wastes

Management of solid biohazardous wastes contaminated with radioactive materials is a potentially serious problem for containment facilities because of requirements to inactive biohazardous properties of the waste before it is removed from the containment area (Richmond & McKinney, 1996). Some facility regulations require on-site inactivation of wastes from microbiology laboratories, regardless of the biosafety level (USAGFD, 2001). Inactivation methods for biohazards such as autoclaving may be incompatible with other hazardous constituents in MHW.

A method for safely autoclaving solid wastes contaminated with the human immunodeficiency virus and low levels of radionuclides commonly used in biomedical research (^{35}S, ^{125} I, ^{32}P, tritium, ^{14}C, and ^{51}Cr) has been described (Stinson et al., 1990a, 1990b). The method, which may be applicable to radioactive wastes contaminated with other infectious agents, employed a double polypropylene bag system and absorbents to contain the radioactivity. Wipe and air samples collected inside the autoclave indicated that no radioactivity escaped from the bags. Note that polyethylene bags should not be used even for storage of radioactive wastes prior to autoclaving. They are permeable to gases ^{125}I emitted from aqueous solutions (Tanaka, 1984).

Prion-contaminated Wastes

Prions are the infectious agents associated with a group of neurodegenerative diseases collectively described as transmissible spongiform encephalopathies (TSEs) or transmissible degenerative encephalopathies or (TDEs). They are believed to consist of small proteinaceous particles and do not contain detectible nucleic acids. Prions are extraordinarily resistant to most conventional methods inactivation (Brown et al., 2004a; Taylor, 2000, 2004). Infectivity may remain ash from infected tissues heated to 600°C (Brown et al., 2000, 2004a, 2004b). Laboratories anticipating research on TDEs should ensure that they have capability to dispose of prion-contaminated wastes by one of these options:

- Immersion in either 1N sodium hydroxide or sodium hypochlorite (20,000 ppm available chorine) for at least one hour before autoclaving at 134°C for 4.5 hours,
- Hot alkaline hydrolysis and
- Incineration.

Management of Contaminants in Sterilization System Tankage

Guidelines published in Biosafety in Microbiological and Biomedical Laboratories (Richmond & McKinney, 1999) and facility regulations (USAGFD 2001) require inactivation of contaminated wastewater from high containment laboratories before it is discharged to the sanitary sewer. This is usually accomplished by collecting the wastewater from biocontainment areas in tanks and sterilizing it with heat. Because wastewater from tank systems is considered to be a nondomestic or industrial waste discharge in many jurisdictions, it is often subject to more stringent monitoring requirements and discharge limitations than wastewater continuously discharged to sanitary sewage systems (USAGFD, 1996; WSSC, 1995). This raises significant compliance issues:

- Contaminant levels and other regulated parameters (e.g., temperature, pH, and biological oxygen demand) are measured in the tankage, not after blending with other wastewater in the sewerage system, which equalizes and may dilute contaminants below discharge limits.
- Discharge regulations and operating permits may require each batch to be analyzed and the results of the analysis reported to the receiving wastewater treatment facility.
- Time consuming regulatory agency discharge approvals may be required before the batch can be released to the sewer.
- Larger wastewater holding capacity may be required to allow time for batch analyses and securing discharge approvals.

Wastewater contaminant issues for the various types of contaminants are presented in the following sections.

Radioactive Contaminants

Most USNRC radioactive materials licenses and facility waste handling guidance (NIH, 2004) require radioactive wastes to be put in containers, collected, and disposed of through centralized services rather than being discharged directly to the sanitary sewer. However, radioactive contaminants could potentially be present in wastewater from containment facilities where protocols involving administration of radioactive materials to animals are carried out. These contaminants may be present in excreta, necropsy wastes or other non-containerized wastes that are discharged to the wastewater system.

NRC regulations exempt excreta from individuals undergoing medical diagnosis or therapy from record keeping and other requirements applicable to radioactive wastes.[13] The two conditions for exemption that must be satisfied are: (1) the matter to be disposed of must be excreta, and (2) the excreta must be obtained from individuals undergoing medical diagnosis or therapy with radioactive materials. The exemption applies even when disposals of patient excreta do not follow direct routes from patient to sewer (e.g., urine samples sent to a laboratory for analysis). *It is important to emphasize that this exemption does not apply to excreta from laboratory animals or wastes from procedures such as necropsy.*

---

[13] 10 CFR 20.303 (d) [or 10 CFR 20.2003 (b)]

Radioactive contamination of wastewater should be prevented by:

- Limiting and isolating the areas where radioactive materials are used.
- Collecting potentially contaminated wastes in containers
- Surveying liquid wastes before discharge
- Preventing use or disposal of radionuclides in activity concentrations that could exceed discharge requirements
- Holding wastewater contaminated with short-lived radionuclides for decay

Some contamination of wastewater is probably unavoidable. Techniques for removal of radionuclides in wastes from a few biomedical procedures such as radioimmunoassay have been developed to minimize discharges of radioactive materials in storage tanks (Tsuchiya et al., 1983) but they are probably not practical for treatment of large volumes of tankage.

Installation of multiple holding tanks in wastewater systems is a potential strategy to deal with wastewater contaminated with short-lived radionuclides. Neung and Nikolic (1998) used such tanks to increase the holding time of excreta from thyroid cancer patients treated with ^{131}I. This provided more time for physical decay before discharge of the wastewater to the local sewerage system. Larger patient loads could be accommodated without exceeding permissible activity concentrations. This same strategy could be applied to the design of biomedical research laboratories to increase the number of protocols that can use short-lived radionuclides and may generate radioactive wastewater.

Toxic Organic Contaminants

Laboratory facility compliance with industrial wastewater requirements for control of toxic chemical contaminants may require enormous effort and costly monitoring and treatment systems. NIH experience with discharges from tank systems used to decay radioactive waste water is illustrative.

The NIH is serviced by the Washington Suburban Sanitary Commission (WSSC) which provides water and wastewater disposal service to the installation and regulates nondomestic discharges. WSSC regulations and discharge authorization permit conditions require the NIH to test each batch of decayed wastewater and obtain approval before it is released to the sanitary sewer, which occurs several times a year. Testing requirements include a flash point determination, pH, and Total Toxic Organics (TTO)[14], which is the sum of concentrations for approximately 100 volatile and semivolatile organic compounds listed as priority pollutants. Each compound must be determined at a minimum detection limit of 10ppb which requires complex extraction procedures and gas chromatography/mass spectroscopy (GS/MS) instrumentation. To assure that the analytical results are received in time to meet the requirements of tank processing and clearance, the analyses must be performed in a dedicated, on-site laboratory. TTO compounds frequently detected in untreated wastewater, which, based on NIH experience, are likely also to be present in wastewater from containment facilities include:

- Chloroform as a reaction product of sodium hypochlorite disinfectants and organic materials
- Phthalate plasticizers, particularly di-n-butylphthalate which may leach from polyvinylchloride plastics
- Phenol from extraction procedures
- Phenolic compounds, as components, contaminants, or breakdown products of disinfectants

It should be understood that the wastewater tanks at NIH are not heated. Other toxic organic compounds could potentially formed as degradation products or recombinants from heating wastewater and chemical disinfectants in containment facility cook tanks that were not found in wastewater from the radioactive waste decay tanks at NIH. Another concern is the potential for toxicity to biological treatment systems at wastewater treatment facilities which could arise from the disinfectants present in large volumes of wastewater generated by use of chemical showers used in high containment facilities.

The need to control toxic organic compounds in wastewater has significant potential impacts on facility design and operation. Some strategies that should be considered by developers and operators of containment facilities include:

1. Avoiding use of materials in facility infrastructure and equipment that may be sources of contaminants. For example, phthalate contamination may be preventable by avoiding use of polychlorinated vinyl plastics (PVC) in equipment and materials that will be in contact with wastewater such as laboratory tubing; sanitary waste system piping; flooring and wall coverings in areas that require cleaning and disposal of rinsates into the wastewater system.

2. Selecting disinfectants and conditions of their use to minimize wastewater volume and toxicity.

3. Working proactively with local wastewater authorities to develop operating plans that assure regulatory compliance and protection of biological

---

[14] The WSSC TTO discharge limit is 2.13 mg/l.

treatment systems but do not impose unnecessarily burdensome or infeasible monitoring analysis requirements. It will not be feasible to conduct the same level of detailed sampling and analysis that NIH performs on cook tank effluents which will probably have to be discharged from containment facilities on a frequent, perhaps daily schedule. An approach may be to collect and analyze a full set of reference samples from a standardized treatment process for validation and then only require additional sampling and analyses whenever there have been significant changes to the wastewater treatment and generation processes, such as a change in disinfectants used for chemical showers. This would obviate the need to develop costly on-site laboratory capability for TTO analysis.

4. Consider the potential need to treat wastewater from cook tanks prior to discharge to the sanitary sewer. The NIH ensures compliance with the local TTO limit by filtering the contents of containers of aqueous radioactive wastes collected from laboratories with granular activated carbon before it is discharged into the tanks for decay and subsequent discharge to the sewer.

Summary

Mulithazardous waste streams generated from biomedical laboratories result in management burdens that are often inordinate to their small volume and the relative risks they pose to safety and the environment. For high containment facilities these burdens will be reduced by ensuring that all investigators receive effective training on MHW minimization and management, and incorporating features into designs and operating plans that will minimize generation of problematic waste streams and maximize capability for on-site treatment and disposal. This will reduce the amount of wastes that must be shipped off-site, reducing costs, enhancing facility sustainability and the security of the research mission.

While the overall environmental impact of increased MHW generation from containment facilities will probably be negligible, methods and technologies developed to minimize waste generation, decontaminate and dispose of unavoidable MHW in a manner that is safe, protective of the environment, and cost effective may have broader applications in health care, and improving the Nation's capability to respond to future bioterrorism threats and disease outbreaks.

References

ACS (American Chemical Society). (2004). Less is better: Guide to minimizing waste in laboratories. Available at: http://membership.acs.org/c/ccs/pubs/less_is_better.pdf.

ATSDR (Agency for Toxic Substances and Disease Registry). (2004). Toxicological profile for mercury) (March 1999). Available at: http://www.atsdr.cdc.gov/toxprofiles/tp46.html.

Austin. S.M., Rau, E.H, Holcomb, W.F., & Zoon R.A. (2002). Reduction in radioactive material use and waste generation at the National Institutes of Health. *Health Physics*, **83**(11), S85-95.

Bagenstose, L.M., Mentink-Kane M.M., Brittingham, A., Mosser, D.M., & Monestier M. (2001). Mercury enhances susceptibility to murine leishmaniasis. Parasite Immunology, **23**(12), 633-640.

Barker, L.F., Rau, E.H., Pfister, E.A., & Calcagni, J. (2000). Development of a pollution prevention and energy efficiency clearinghouse for biomedical research facilities. Environmental Health Perspectives, **108**(6), 949-951.

Brown, P., Rau, E.H., Johnson, B.K, Bacote, A.E, Gibbs, C.J., Jr, & Gajdusek, D.C. (2000). New studies on the heat resistance of hamster-adapted scrapie agent: Threshold survival after ashing at 600 degrees C suggests an inorganic template of replication. Proceeding of National Academy of Sciences USA, **97**(7), 3418-3421.

Brown, P., Rau, E.H., Meyer, R, Lemieux, P., Cardone. F, & Pocchiari, M. (2004a). Extreme inactivation methods for transmissible spongiform encephalopathy agents. 8th International Kilmer Memorial Conference, Osaka, Japan. *International Kilmer Conference Proceedings*, **8**, 104-111.

Brown, P., Rau, E.H., Lemieux, P., Johnson, B.K., Bacote, A.E., & Gajdusek D.C. (2004b). Infectivity studies of both ash and air emissions from simulated incineration of scrapie-contaminated tissues. Environmental Science and Technology, **38**(22), 6155-6660.

CDC (Centers for Disease Control and Prevention). (2004). Blood mercury levels in young children and childbearing-aged women — United States, 1999—2002. MMWR, **53**(43), 1018-1020.

Clinton, W.J. (1998). Executive Order 13101 of September 14, 1998. Greening the Government through Waste Prevention, Recycling, and Federal Acquisition. Federal Register, **63**(179), 49643-49651.

Clinton, W.J. (2000). Executive Order 13148 of April 21, 2000. Greening the Government through Leadership in Environmental Management. Federal Register, **65**(81), 24595-24606.

Costello, R.G., Emery, R.J., Pakala, R.B., & Charlton, M.A. (2000). Radioactive waste minimization implications of clinically-indicated exsanguination procedures. Health Physics, **79**(3), 291-293.

Gartner, L. (2004). Security issues related to designing high-containment facilities. Applied Biosafety, **9**(3), 155-159.

Goddard, C. (1999). The use of delay tanks in the management of radioactive waste from thyroid therapy. Nuclear Medicine Communications, **20**(1), 85-94.

H2E (Hospitals for a Healthy Environment). (2004). Tools/resources: Mercury. Available at: http://www.h2e-online.org/tools/mercury.html.

Hageman, J.P. (2002). Handling, storage, treatment, and disposal of mixed wastes at medical facilities and academic institutions. Health Physics, **82**(5), S66-76.

Holcomb, W.F. (1995). Radiation safety training at the National Institutes of Health. Military Medicine, **160**(3), 115-120.

Holcomb, W.F., Austin, S.M., Rau, E.H., Zoon, R.A., & Walker, W.J. (1993). Training as a strategy for management of biomedical mixed waste at the National Institutes of Health. Seattle, WA: Summer Meeting, American Institute of Chemical Engineers.

Kaye, G.I., & Weber, P.B. (1992). Method for disposing of radioactively labeled animal carcasses. U.S. Patent 5,332,532. U.S. Patent and Trademark Office, Washington DC.

Kaye G, Weber P, Evans A, Venezia R. (1998). Efficacy of alkaline hydrolysis as an alternative method for treatment and disposal of infectious animal waste. Contemporary Topics in Laboratory Animal Science 1998; **37**(3), 43-46.

Kaye, G.I., & Weber, P.B. (2004). Treatment and disposal of biological, biohazardous, and regulated medical waste by alkaline hydrolysis: The WR^2 process. Available at: http://www.yale.edu/oehs/hwc1999/hwc/kaye.pdf.

Leung, P.M., & Nikolic, M. (1998). Disposal of therapeutic 131I waste using a multiple holding tank system. *Health Physics,* **75**(3), 315-321.

MASCO (Medical Academic and Scientific Community Organization, Inc.) (2004a). Mercury work group site index. Available at: http://www.masco.org/mercury/index.html.

MASCO (Medical Academic and Scientific Community Organization, Inc.). (2004b). Mercury database. Available at: http://www1.netcasters.com/mercury/.

(MPCA) Minnesota Pollution Control Agency. (2004). Lake Superior Basin Water Standards. Chapter 7052.0260. Compliance Schedules. Available at: http://www.revisor.leg.state.mn.us/arule/7052/0260.html.

NCCLS (National Council for Clinical Laboratory Standards). (2002). Clinical laboratory waste management: Approved guideline — 2nd ed.) GP5-A2. Wayne, PA: National Committee for Clinical Laboratory Standards, Inc.

NCRP (National Council on Radiation Protection and Measurements). (2003). Management techniques for laboratories and other small institutional generators to minimize off-site disposal of low-level radioactive waste. NCRP Report No. 143, Bethesda, MD: National Council on Radiation Protection and Measurements.

NIH (National Institutes of Health). (1992). Mixed waste: A major issue in biomedical research (A video film). Bethesda, MD: National Institutes of Health.

NIH (National Institutes of Health). (1995). Report of the NIH Committee on Alternatives to Radioactivity (NCAR) to the Deputy Director of Intramural Research. Bethesda, MD: National Institutes of Health.

NIH (National Institutes of Health). (1995). Pollution prevention plan for the National Institutes of Health. Bethesda, MD: Environmental Protection Branch. Division of Safety, National Institutes of Health.

NIH (National Institutes of Health). NIH initiative to reduce regulatory burden. Identification of issues and potential solutions. 1999. Available at: http://grants2.nih.gov/grants/policy/regulatoryburden/. Accessed 2004.

NIH (National Institutes of Health). NIH Mercury Abatement Program. Available at: http://orf.od.nih.gov/nomercuryhome.htm. Accessed online 2004.

NIH (National Institutes of Health). NIH waste disposal guide. Available at: http://orf.od.nih.gov/waste/wasteguide.html. Accessed online 2004.

NRC (National Research Council). Prudent Practices in the Laboratory: Handling and Disposal of Chemicals. National Academy Press, Washington, DC, 1995.

ODH (Ohio Department of Health). Suggested action levels for indoor mercury vapors in homes. Available at: http://www.odh.state.oh.us/odhprograms/hlth_as/MercPack/Indoor.pdf. Accessed 2004.

OSHA (Occupational Safety and Health Administration). Standard Interpretations 09/03/1996 - PEL for inorganic mercury is a time weighted average, not a ceiling. Available at: http://www.osha.gov/pls/oshaweb/owadisp.show_document?p_table=INTERPRETATIONS&p_id=23866. Accessed online 2004.

Party, E., & Gershey, E.L. (1995). A review of some available radioactive and non-radioactive substitutes for use in biomedical research. Health Physics, **69**(1), 1-5.

Rau, E.H. (1997). Trends in management and minimization of biomedical mixed wastes at the National Institutes of Health. Technology, Journal of the Franklin Institute, **334**(A), 397-419.

Rau, E.H. (2000a). Mad as a Hatter? The campaign for a mercury-free NIH. Paper 10, Division of Chemical Health and Safety Technical Session: The Health Hazards of Mercury. New Orleans, LA: 225th American Society National Meeting.

Rau, E.H. (2003b). Assessment and removal of mercury contamination in laboratory decommissioning activities. Paper 11, Division of Chemical Health and Safety Technical Session: The Health Hazards of Mercury. New Orleans: 225th American Chemical Society National Meeting.

Rau, E. (2004). NIH going mercury free. FedFacs, **16**(10).

Rau, E.H., Alaimo, R.J., Ashbrook, P.C., Austin, S.M., Borenstein, N., Evans, M.R. et. al. (2000). Minimization and management of wastes from biomedical research. Environmental Health Perspectives, **108**(6), 953-977.

Rau, E.H., & Carscadden, J. (2004). Successful implementation of the new NIH facility decommissioning protocol: Results of a pilot project. Bethesda, MD: 3rd Annual DHHS Environmental Networking Training Workshop.

Richmond, J.Y., & McKinney, R.W., (Eds). (1999). Biosafety in microbiological and biomedical laboratories. (4th. Ed). Washington, DC: U.S. Government Printing Office.

Richmond, J.Y, Hill, R.H., Weyant, R.S., Nesby-O'Dell, S.L., & Vinson, P.E. (2003). What's hot in animal biosafety. ILAR Journal, **44**(1), 20-27.

Salako, Q., & DeNardo, S.J. (1997). Analysis of long-lived radionuclidic impurities in short-lived radiopharmaceutical waste using gamma spectrometry. Health Physics, **72**(1), 56-59.

Shen, X., Lee, K., & König, R. (2001). Effects of heavy metal ions on resting and antigen-activated CD4(+) T cells. Toxicology, **169**(1), 67-80.

Silbergeld, E.K., Sacci, J.B., Jr., & Azad, A.F. (2000). Mercury exposure and murine response to Plasmodium yoelii infection and immunization. Immunopharmacology and Immunotoxicology, **22**(4), 685-695.

Stinson, M.C., Galanek, M.S., Ducatman, A.M., Masse, F.X., & Kuritzkes, D.R. (1990). Model for inactivation and disposal of infectious human immunodeficiency virus and radioactive waste in a BL-3 facility. Applied and Environmental Microbiology, **56**(1), 264-268.

Stinson, M.C., Galanek, M.S., Ducatman, A.M., Masse, F.X., & Kuritzkes, D.R. (1990). Model for inactivation and disposal of infectious human immunodeficiency virus and radioactive waste in a BL-3 facility. Applied and Environmental Microbiology, **56**(1):264-268.

Stinson, M.C., Green, B.L., Marquardt, C.J., & Ducatman, A.M. (1991). Autoclave inactivation of infectious radioactive laboratory waste contained within a charcoal filtration system. Health Physics, **61**(1), 137-142.

Sustainable Hospitals-Lowell Center for Sustainable Production. (2004). Mercury reduction. Available at: http://www.sustainablehospitals.org/HTMLSrc/IP_Mercury.html

Sweet, L.I., & Zelikoff, J.T. (2001).Toxicology and immunotoxicology of mercury: A comparative review in fish and humans. Journal of Toxicololgy and Environmental Health B: Critical Reviews, **4**(2), 161-205.

Tanaka, Y. (1984). [Leakage of gaseous 125I from a Na125I solution through a plastic film]. Radioisotopes, **33**(9), 629-31. [Article in Japanese]

Taylor, D.M. (2000). Inactivation of transmissible spongiform encephalopathy agents: A review. Veterinary Journal, **159**, 10-17.

Taylor, D.M. (2004). Inactivation of prions. 8th International Kilmer Memorial Conference, Osaka, Japan. International Kilmer Conference Proceeding, **8**, 93-103.

Tchounwou, P.B., Ayensu, W.K., Ninashvil, N., & Sutton, D. (2003). Environmental exposure to mercury and its toxicopathologic implications for public health. Environmental Toxicology, **18**(3), 149-175.

Tsuchiya, T., Norimura, T., & Ueno, T. (1983). A system for decontamination of liquid radioactive waste produced in in vitro tests in nuclear medicine. Radioisotopes, **32**(6), 286-288. [Article in Japanese]

USAGFD (U.S. Army Garrison at Fort Detrick, Department of the Army). (1996). Management of medical waste. Environmental Quality. Nondomestic Wastewater Control, Fort Detrick Regulation 200-7. Fort Detrick, MD: U.S. Army Garrison.

USAGFD (U.S. Army Garrison at Fort Detrick, Department of the Army). (2001). Management of medical waste. Fort Detrick Regulation 385-4. Fort Detrick, MD: U.S. Army Garrison.

USDOE (U.S. Department of Energy). (1994). National Institutes of Health: Mixed waste stream analysis. National Low Level Waste Management Program. DOE/LLW-208. Idaho Falls, ID: U.S. Department of Energy.

USDOE (U.S. Department of Energy). (1995). National Institutes of Health: Mixed waste minimization and treatment. National Low Level Waste Management Program. DOE/LLW-218. Idaho Falls, ID: U.S. Department of Energy.

USEPA (U.S. Environmental Protection Agency). (1986). Federal Register, **51**, 10168.

USEPA (U.S. Environmental Protection Agency). (2004). Integrated Risk Information System. Mercury, elemental (CASRN 7439-97-6). File first online 1988. Available at: http://www.epa.gov/iris/subst/0370.htm#refinhal.

USEPA (U.S. Environmental Protection Agency). (2000). Environmental management guide for small laboratories EPA 233-B-00-001. Washington, DC: Office of the Administrator, U.S. Environmental Protection Agency.

USEPA (U.S. Environmental Protection Agency). (2001a)...Storage, treatment, transportation, and disposal of mixed waste. Federal Register, **66**(95), 27218-27266.

USEPA (U.S. Environmental Protection Agency). (2004) Environmental fact sheet: Low-level mixed waste conditionally exempt from hazardous waste regulation. EPA530-F-01-008. 2001b. Available at: htpp://www.epa.gov/radiation/mixed-waste/docs/factsheet.pdf.

USEPA (U.S. Environmental Protection Agency). (2003). Approaches to an integrated framework for management and disposal of low-activity radioactive waste: request for comment; proposed rule. Federal Register, **68**(222), 65120-65150.

USEPA (U.S. Environmental Protection Agency). (2004). Mercury pollutant minimization program guidance. U.S. Environmental Protection Agency, Region 5, NPDES Programs. Available at: http://www.epa.gov/region5/water/npdestek/npdprta.html.

USNRC (U.S. Nuclear Regulatory Commission). (1981). Biomedical waste disposal — Nuclear Regulatory Commission. Final rule. Federal Register, **1146**(47), 16230-16234.

USNRC (U.S. Nuclear Regulatory Commission). (2004). NRC Regulatory issue Summary 2004-7: Revised decay-in-storage provisions for the storage of radioactive waste containing byproduct material. Available at: http://www.nrc.gov/reading-rm/doc-collections/gen-comm/reg-issues/2004/ri200417.pdf.

WSSC (The Washington Suburban Sanitary Commission). (1998). Regulations governing industrial waste in the Washington Suburban Sanitary District. Revised 1995. Available at: http://www.wssc.dst.md.us/rsg/IndustrialDischarge/forms/Chapter9.pdf .

Wolfram, J.H., Rogers, R.D., Wey. J.E., & Rau, E.H. (1997). Degradation of hazardous chemicals in liquid radioactive wastes from biomedical research using a mixed microbial population. Technology, Journal of the Franklin Institute, **344**(A), 171-184.

World Health Organization. (2004). Laboratory biosafety manual. Geneva : World Health Organization.

Yoder, K.D. (2002). Managing a low-level mixed waste storage facility: A checklist for compliance to 40 CFR 266. Health Physics, **82**(5), S77-81.

Chapter 8

Viral Genetic Diversity Network (VGDN) in Brazil

Edison Luiz Durigon

Abstract

Despite the relative complexity of scientific activity in the state of São Paulo, Brazil, the subject of virology lacks sufficient capacity in both its basic and applied aspects, even considering the widespread activities on viral diagnosis and clinical intervention. The reasons for this are many and range from a lack of a traditional training in virology in the State to a difficulty in establishing groups acting in this multidisciplinary field, which interface several disciplines in the biomedical sciences. This program proposes setting up of a network of laboratories at several research institutions in the state to study the genetic diversity of viruses within an applied science context, which will provide both competence in genetic studies of viruses and relevant primary data of immediate use (e.g., monitoring vaccines, antiviral drug resistance and viral typing for information and disease associations). Several examples of the importance of this initiative will be presented herein. When fully functional, the VGDN should provide services for society via the articulation of coordinated activities under epidemic outbreak conditions. Under ideal conditions the VGDN should be able to operate under real emergency conditions with at least four groups operating at biosafety level 3. The establishment of a Viral Genetic Diversity Network (VGDN) will be germane for the improvement of science and technology in the state of São Paulo, Brazil and will provide much needed support for epidemiological surveillance and the control of viral communicable diseases.

Introduction

Viral Diseases: Examples and Tendencies.

Since the beginning of the 20th century, human virology has become a major area of biomedical investigation after the characterization of the yellow fever virus. The exponential growths of human populations in the last 200 years, followed by poverty, irrational urbanization, military conflagration and environmental degradation, which have been the hallmarks of the 20th century, are probably important conditions for the worldwide dispersion and establishment of communicable infectious diseases. Despite significant technological progress during this century, infectious diseases have been responsible for levels of mortality higher than any natural or man-induced causes.

At the end of the First World War (1914-1918), a pandemic caused by the influenza virus killed around 20 million people worldwide in a few months, exceeding the mortality during the war itself. Although vaccines can readily be obtained for the new variants and reassortants of the flu virus, health authorities everywhere are aware of the high likelihood of another flu pandemic, expected to cause an excess of 60 million causalities some time in the future. Other viruses pose problems too; the hepatitis B virus (HBV) kills on average 1 million people every year. Similar mortality figures are observed for the measles virus, while the human immunodeficiency virus (HIV), currently infects tens of millions of persons. In certain African nations (e.g., Botswana), around 40% of the adult population are carriers of HIV, and the incidence of the virus appear to be increasing in developing and poor nations (e.g., Brazil, Thailand and India among others). Initially highly prevalent among risk groups, such as hemophiliacs, intravenous drug users and male homosexuals, in the last few years HIV has been making its way into adolescent youth, women, children and newborns despite prevention efforts. The Brazilian Government spends hundreds of millions of dollars each year distributing antiretroviral drugs among thousands of carriers. Nevertheless, there is no good information on the prevalence of resistance-conferring mutations to the drugs in patients naive of treatment, nor do we have any idea on the dynamics of propagation of resistance among recent infections.

Other viral diseases, such as dengue, have dispersed across all equatorial and tropical regions since the last war. Dengue currently causes about 150 million infections and 25,000 deaths every year. Until the second half of the 20th century, dengue was considered a benign disease but, possibly assisted by an increase of genetic diversity and to distinct serotype co circulation, two potentially lethal forms of the disease emerged: dengue hemorrhagic fever (DHF) and dengue shock syndrome (DSS). Moreover, it is possible that dengue virus strains from South East Asia, but isolated in the Caribbean early in this decade, could be associated with higher likelihood of DHF or DSS than those circulating in other areas of the world. By observing the epidemiological patterns during the last 50 years, it could be argued that in the near future all serotypes will potentially be co-circulating among the vast majority of geographical areas and that a significantly high proportion of the human population, living on epidemic zones, will have been exposed to one or more co circulating heterologous subtypes and as a consequence suffer severe forms of disease. Due to the recent human population growth, increase in the prevalence of several other viral diseases may also be occurring, such as several other arboviruses (e.g., rift valley fever virus, hepatitis viruses, enteroviruses and several respiratory viruses).

The Importance of the Study of Viral Genetic Diversity

Historical evidence, retrospective studies and viral population genetics studies suggest that human populations are being exposed to an increasing viral genetic diversity. One of the main factors dictating the interaction among virus and human host is the asymmetry in the amount of genetic diversity the former can produce compared to that generated by the immune response of the latter. This can be explained by the fact that RNA viruses show mutation rates one million fold higher than that of the eukaryotic cell DNA. Such rapid accumulation of genetic diversity is the main reason why antiviral drugs and vaccines are so difficult to develop. Detection of variants, specific mutations, and genomic features of viruses are therefore important for disease, prevention, prognosis, and therapy.

Importance of Primary Sequence Data for Therapy

For several human viruses well defined correlations have been established between genetic or genotypic traits and biological characteristics. These phenotypic traits can be relevant for disease prevention and control, making genetic information of viruses a useful measure. For example, the hepatitis C virus (HCV) may carry mutations in its RNA-dependent RNA polymerase, which are associated with interferon resistance. Moreover, the subtype 1b of HCV shows a weaker response to interferon and has been proposed to be correlated with severe hepatic disease. It is therefore evident that knowing the genetic information in HCV is very important for choosing sensible therapeutic strategies for specific patients. Phenotypic information regarding HIV-1 is also directly available from nucleotide sequences. For example, the net charge in the V3 loop of the gp120 of the HIV-1 is associated with tissue tropism and perhaps disease prognosis. Additionally, well-defined mutations in the genes coding for both protease and reverse transcriptase genes of HIV-1 correlate with resistance to antiretroviral drugs, making their detection crucial for deciding on drug combination and rotation strategies.

Importance of Primary Sequence Data for Epidemiological Studies

Having direct genetic information on the viral strains involved in an outbreak is of crucial importance, providing information regarding, (i) where the viruses come from, (ii) information on the possible time of strain introduction, and (iii) on its status as hypo endemic in the affected population. In the case of preventive vaccines being used on a large scale, relevant information on the actual protection provided by the vaccine can be made readily available by sequence data from the circulating viral strains. If attenuated viral strains are being used in vaccination programs, wild type revertant vaccine viruses can be detected and may provide important information concerning the actual progress of an outbreak.

Importance of a Network for the Study of Viral Genetic Diversity.

The appropriate quantitative and qualitative assessment of genetic diversity from viral populations based on sequence data can help us understand the structure of epidemics as it unfolds, allowing for better intervention strategies. This argument stands as the reason for enabling the VGDN to operate in coordinated tasks (see below), which will also allow the gauging of both performance and competence of its member groups. The study of viral genetic diversity is also of fundamental importance for establishing competence in virology and in providing conditions for obtaining fast, direct and relevant information for issues in public health. With rare exceptions, in emergency situations, samples are sent abroad for characterization and analysis. On several instances, publications based on viral isolates from Brazil do not even acknowledge the effort of the crucial initial steps of isolation, which were mainly done by local scientists and technical personnel. Moreover, other than the unnecessary dependence, national interests may also come into play, since genetic information can be relevant for the production of vaccines, implementation of vaccination programs, development of antiretroviral drugs, and estimation of financial risk for different economical activities (urban planning and management, public health budgeting, health-associated industrial cost, insurance company cost benefit evaluations, etc.) and for setting priorities for control and surveillance programs.

Importance of Viral Genetic Diversity Network (VGDN) in Brazil

Hepatitis B virus (HBV) is a relevant example of the need for a better understanding of viral genetic diversity is HBV in Brazil. Other than genotypes A and D, both of which are found in developed countries, there is a high prevalence of the F subtype in indigenous populations in Brazil, and the F subtype appears to be a New World specific virus. Moreover, in the Amazon region, fulminates cases of hepatitis B are associated with co infection with the delta (viroid-like) agent (hepatitis D virus), about which not much is known in Brazil. Based on previous incipient serological studies on the prevalence of HBV, it is believed that 30% of the Brazilian and Peruvian Amazon population have been exposed to the virus. A detailed study on the prevalence of HBV subtypes in Brazil and their detection by the commercially available kits is therefore important and might have an impact in reducing blood transfusion related HBV transmission.

Another important example is HIV-1 in Brazil. The detection of differences in prevalence through time (incidence) of the many HIV-1 subtypes around the World is becoming a much sought after information for formulating meaningful vaccination strategies. Although it is the most prevalent subtype in the U.S. and Europe, subtype B was surpassed by subtype E in Thailand. In Brazil, subtype F (also found in central Africa and Romania), is responsible for more than 20% of recent infections, most of which are among intravenous drug users. The high prevalence of subtype C was detected in some regions in the South of Brazil. Moreover, ongoing studies carried at the UNIFESP (Federal University of Sao Paulo) and UFRJ (Federal University of Rio de Janeiro), suggest that HIV-1 subtypes in Brazil show variants not found elsewhere, possibly associated with efficient transmission early in infection. Moreover, the variability in the variable loop in the gp120 of HIV-1 in the blood donors in the Amazon is quite high compared to other regions, which cannot be explained simply by the numbers of infected people in that region. HIV-1 research in Brazil has been receiving considerable support from the Federal Government, partly because of an intense social demand, voiced by nongovernmental organizations directly to the Ministry of Health. Unfortunately, such concerted effort is not exerted on behalf of the control of other viral agents, which are main factors keeping high infant mortality rates (e.g., enteroviruses, rotaviruses, etc.).

A further example which stands as a justification for setting the VGDN is the measles virus. This virus was considered under control in the early 1980's in the USA. Today measles causes 1 millions deaths worldwide. Outbreaks were observed in vaccinated populations for reasons not well understood which certainly give this virus a capacity to generate genetic diversity. In Brazil, a major outbreak was recently observed. The increase in notified cases from 22 in 1996 to 70.000 in 1997, with several deaths, makes this virus a serious urban public health problem, given its easy transmission, which is known to be directly related to population density and size. In spite of these facts, studies on the molecular epidemiology of this virus were carried in the CDC in Atlanta. The results of this study will certainly reveal important information on the nature of the viral strains involved in the outbreak and on the vaccines being used along 1996 and 1997.

Coordinated Tasks and Emergencies

When fully functional, the VGDN should provide services for society via the articulation of coordinated activities under epidemic outbreak conditions. Because of its response agility, technical capabilities need to be tested continuously. This will allow for a mechanism for Fundação de Amparo a Pesquisa do Estado de São Paulo (FAPESP) to gauge the performance of each individual group and their combined operation capabilities. An initial role of the coordinated tasks would then allow for a mechanism of accountability and for monitoring performance. They also constitute a unifying element of the VGDN, offering a concrete operational justification. The coordinated tasks will generate relevant primary data, given the judicious choice of agents to be studied. During its operation, several agents may be tackled (e.g., dengue, Hantavirus, Arbovirus, RSV, measles, viral hepatitis, rotavirus, HIV-1, etc.), however, the choices should be a function of the level of competence reached by the VGDN. Therefore the tasks should present increasing difficulty along time. Under ideal conditions the VGDN should be able to operate under real emergency conditions. It should not escape our attention that a nimble mechanism for supplying reagents and other materials has to be in place during emergencies. It is also true that an agile system for importation of reagents is not at place in the state at this moment. In sum, coordinated tasks are a mechanism for reaching high competence levels, assessment and detection of deficiencies.

Objectives

General Objectives of the VGDN

The network aims to empower several laboratories to do molecular epidemiology and to be able then to operate in a coordinated manner in emergency situations. It will function as a multicenter, multitask network. Member groups will not need to be working on the same virus and will be encouraged to pursue their own independent line of research in order to broaden the scope of agents being studied by the network. However on special occasions, all nodes will have to be able to cooperate on coordinated tasks. Among the unifying elements and tasks for the VGDN are:

(i) A system of sampling and experimental design based on clinical and epidemiological information, which will maximize the utility of the genetic information.

(ii) The study of the genetic diversity by means of large scale automated sequencing of informative regions of several different virus genomes. This will generate sequence data about viral genetic diversity of use for epidemiological surveillance and control.

(iii) Follow up genetic variants during epidemic outbreaks and assist in the prediction of possible outbreaks and detect variants of importance for diagnosis and vaccines.

(iv) A high performance bioinformatics system for primary data storage and analysis.

(v) Groups operating at biosafety level 3 (BSL-3). These laboratories can demonstrate capacity and scientific production in virology of level 3 agents.

(vi) Allow for the appearance of more centers dedicated to viral surveillance in São Paulo (e.g., dedicated to emergent diseases) and for a greater number of groups of excellence on molecular virology and capability to handle level 3 agents.

Implementation and Functional Description of the Network

Group Formation

Although sequencing viral genomes will be the main technical goal in this program, it must be brought to mind that sequencing is not going to be the limiting factor. Given that automation is causing DNA sequencing to become less demanding, the limiting factor lays at the experimental design and data analysis steps. Therefore, several training courses will be scheduled to provide technical and conceptual background for researchers, students and technicians, involving the following subjects:

a. Experimental design and sample collection: (i) clinical aspects, (ii) basic and applied epidemiology, and (iii) statistical basis for experimental design and sampling.

b. Basic concepts on molecular epidemiology and modern molecular virology: (i) sample handling, (ii) virus isolation and cultivation, (iii) handling biosafety level 3 agents, and (iv) analysis on genotypic-phenotypic correlations.

c. Data analysis: (i) sequence manipulation and editing, (ii) sequence alignment, (iii) principles of molecular systematic and phylogenetic analysis, and (iv) application of comparative methods: concepts and applications.

Categories of Participant Groups as a Function of the Dynamics of the VGDN

a. Groups operating at biosafety level 1 (BSL-1). Twelve groups without experience in virology (level 1 group) should be linked to sponsor groups. BSL-1 groups can link to BSL-2 or BSL-3 groups according to their study subject. One of the aims of the VGDN is that, in time, groups operating at BSL-1 should migrate to other operational levels in a 4 year period. Groups operating at BSL-1 do not need to cultivate viruses.

b. Groups operating at biosafety level 2 (BSL-2). Seven groups under this designation will work preferably with their own agents already under investigation. Their continued liaison with the VGDN entails the successful participation in coordinated tasks. At this level, groups do not need to have expertise on viral cultivation but, if necessary, should be able to obtain this capability immediately from the technical support of the VGDN. Groups entering the VGDN at BSL-2 are expected to have researchers with a demonstrated competence in virology, as indicated by publications in refereed journals. Viral agents that do not need or are not cultivated in cell lines (HIV-1, HBV, HCV, HGV, etc.) can be studied at biosafety levels 1 or 2.

c. Groups operating at biosafety level 3 (BSL-3). This level entails the operation of four laboratories with BSL-3. These groups can demonstrate installed capacity and scientific production in virology of BSL-3 agents and the physical laboratory requirements consists of the standard BSL-3 requirements as described in the *WHO Laboratory Biosafety Manual* with additional specific requirements for VGDN BSL-3 in Brazil with the following modifications:

i) The laboratory is separated from the areas that are open to unrestricted traffic flow within the building. Entry for personnel through all access spaces is through a double door vestibule with an area designed for protective clothing. Exit from the laboratory is provided with a walk-through shower and a door on the containment barrier (i.e. between the dirty and clean change anterooms). Access doors are self-closing and interlockable (clean and dirty change room doors shall not be opened simultaneously with either the containment laboratory door or the clean change entry door interlock, visual or audible alarms, or protocols are all acceptable). A break-through panel is provided for emergency exit use.

ii) The laboratory room is sealable for decontamination. Air-ducting systems are constructed to permit gaseous decontamination. The building ventilation system is constructed so that air from the laboratory is not re-circulated to other areas within the building. Two exhaust air systems working in backup situation included two fans working 50% of total capacity each and a double bag-in/bag-out type filter with a standby system including two serial HEPA filters each. Visual pressure differential monitoring devices and/or directional airflow indicators were provided at entry to containment barrier. Supply air is interlocked (i.e. fans, dampers, electrical) with exhaust air system, to maintain negative pressure and prevent sustained laboratory positive pressurization.

iii) Double-door barrier autoclave with bio seal is located on containment barrier; and the body of the autoclave is located inside the containment area for safety of maintenance. The barrier autoclave is equipped with interlocking doors, audible alarms to prevent both doors from opening at the same time. The clean side door cannot be opened unless a complete decontamination cycle has been confirmed (time and temperature). The autoclave condensate drain will have a protected system that includes heat flashing of all purge air and liquids prior to the autoclave achieving kill temperature. Drain lines and associated piping (including autoclave condensate and shower) are separated from areas of lower containment and effluents are decontaminated by a heat treatment system (effluent sterilization system) before being discharged to the sanitary sewer. Class II biological safety cabinets are connected to the laboratory exhaust system. The laboratory facility is commissioned and certified annually.

Possible Developments After VGDN

Coordinated tasks will be dictated according to the Public Policies program, which will maintain the operation of most of the nodes, preferably in a network fashion, possibly under emergency conditions. At that time some groups participating at BSL-2, should be entitled to migrate to BSL-3. Level 3 pathogens, needing specialized cultivation will be studied at BSL-2 or BSL-3.

Funding

The implementation of the VGDN will be funded by FAPESP, a Brazilian Foundation from São Paulo State.

Minimal Laboratory Requirements for BSL-1

Twelve groups participating at biosafety level 1 (BSL-1). It consists of equipment for: (i) PCR (automated cycling machine, US$ 8.5K, kits for amplification, US$ 10K), (ii) cloning (shaker, incubator, hood, microfuges and DNA isolation kits, US$ 20K), (iii) sequencer (US$ 100K), (iv) sequencing reagents (US$ 18K, for a total of 200K for 2 years, including sequencing kits and other sequencing-related reagents), and, (v) bioinformatics. (UNIX workstation with video conference capabilities and printer, US$ 10K) (Total of US$ 166.5K).

Minimal Laboratory Requirements for BSL-2

Items of the minimal package (US$ 166K) plus additional resources for viral cultivation and operation at BSL-2, (i) primary barriers (disposable items, such as, gloves, aprons, pipettes, flasks and plastic ware, US$20K), (ii) Class II biosafety hood (US$ 10K), (iii) tissue culture for viruses and or alternative cultivation conditions (e.g., CO_2 incubators US$ 5K, phase contrast microscope, US$ 9K, animals, embryonated eggs, etc., US$ 6K), (iv) stocking systems for cells, viral strains and biological samples (i.e., liquid nitrogen tanks US$ 6K, deep freezer -70^0 US$ 9K and 2 refrigerators, US$ 1.5K. Optional: Serology and immunologicals for the preliminary identification of viral agents (ELISA and immuno-fluorescence system US$ 24K). (US$ 90.5K plus US$ 166.5K, a total of US$ 257K).

Minimal Laboratory Requirements for BSL-3

Items of the minimal package plus conditions of operation at BSL-3. A cost of US$ 350K was estimated for the modification of existing laboratory facilities to operate at the BSL-3 level. This figure considered the need for a negative pressure room (US$ 300K for building modifications and for exhaust air using HEPA filters and refrigeration systems), horizontal autoclaves (US$ 30K) and a dedicated Class II biosafety cabinet (US$ 20K). Therefore the total cost of US$ 697.5K includes US$ 257K for a level 2 laboratory plus the additional US$ 90.5 K needed to duplicate the level 2 functions within the BSL-3 level and US$ 350K for the BSL-3 level facilities. (Total of US$ 697.5K)

Package for Core High Powered Computing Facilities

Basic data analysis (including sequencing editing, alignment and basic polymorphism detection and phylogenetic inference) will be done at each node at a cost of US$ 10K per node. All primary data generated by the network will not be accessed on line at real time by any other part but the member nodes of the VGDN. However, high performance computation will be accessible to all nodes by linking to a parallel machine which will be localized in a bioinformatics laboratory, which will allow global phylogenetic reconstruction and genealogical based inference methods using maximum likelihood models applied to large datasets (correlated changes analysis, selection detection and population parameter estimations such as the rate of growth, etc.). The bioinformatics core unit will also implement a web page for allowing a better integration of the network. In order to preserve privacy and to prevent undesired first hand access to sensible primary data (i.e., clinical and sequence data) the VGDN computer network will be "air-tight". To allow troubleshooting from suppliers of commercial software, selected data will only be transferred by external devices. In order to deal with computation intensive numerical problems arising from sequence data analysis, a 16 CPU parallel machine (or an equivalent Beowulf system) and ancillary equipment is estimated to cost around US$ 350K. Therefore the bioinformatics in the VGDN will consist of one parallel machine (US$ 350K) plus 20 UNIX workstations (US$ 200K, this value was included in the BSL-1 package).

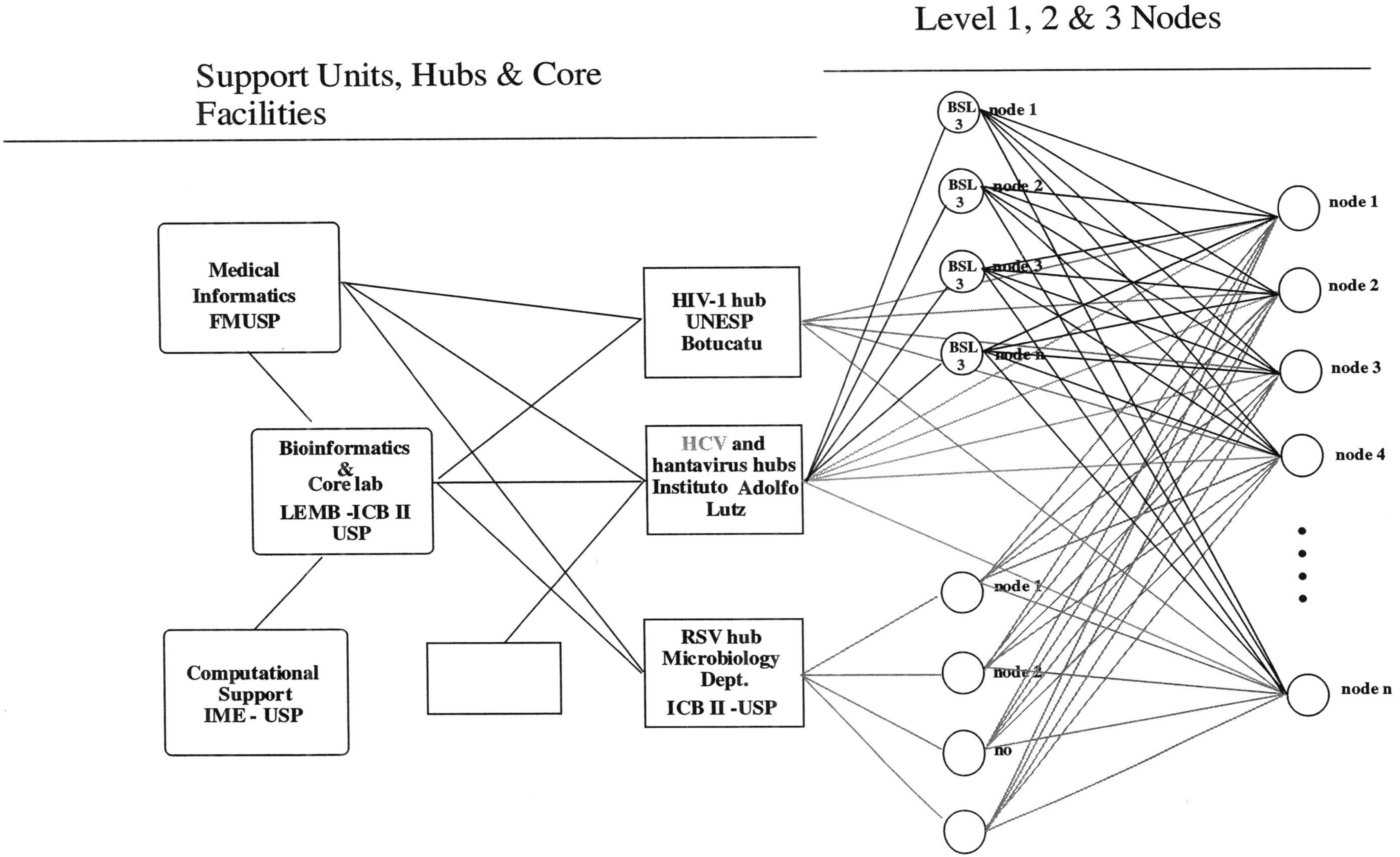

Support Units, Hubs & Core Facilities
Level 1, 2 & 3 Nodes
Medical Informatics FMUSP
Bioinformatics & Core lab LEMB -ICB II USP
Computational Support IME - USP
HIV-1 hub UNESP Botucatu
HCV and hantavirus hubs Instituto Adolfo Lutz
RSV hub Microbiology Dept. ICB II -USP
BSL 3
node 1
BSL 3
node 2
BSL 3
node 3
BSL 3
node n
node 1
node 2
no
node 1
node 2
node 3
node 4
node n

Chapter 9

Working Safely with the Transmissible Spongiform Encephalopathies

Ellen J. Elliott, Claudia MacAuley, Deanna S. Robbins, and Robert G. Rohwer

Introduction

Working safely with the transmissible spongiform encephalopathies (TSEs), also known as "prion" diseases, requires special procedures which are unique to these agents and which differ in substantive ways from containment and decontamination procedures used for most other pathogens. Special handling is needed because of the long-term hardiness of the agent and the difficulty in monitoring for contamination. For example, the human immunodeficiency virus (HIV) if spilled onto a laboratory floor will lose infectivity overnight, while TSE infectivity spilled onto a laboratory floor could be present and infectious 100 years later if not correctly removed. Trace amounts of radioactivity spilled onto the bench top can be detected by a wipe test or a Geiger counter, while trace amounts of spilled TSE infectivity can only be detected by bioassay, which would involve wiping, extracting the wipe, inoculating dozens to hundreds of animals, and holding them for a year or more. Adherence to TSE-specific procedures is required for biosafety reasons, and to preserve experimental integrity by preventing cross-contamination. In this chapter, we describe the procedures that our laboratory has developed for working safely with the TSE agents, and give some perspective on the rationale behind the procedures.

TSE diseases are fatal neurodegenerative diseases that are characterized by vacuolation of the brain and progressive loss of motor and cognitive function. They have been termed unconventional because they have unusually long incubation times (up to decades in humans), because the infectious agents are exceptionally stable and difficult to disinfect by chemical and physical methods, and because the etiologic agent has not been conclusively identified and is proposed to be non-viral and non-bacterial. A prevailing hypothesis is that a self-replicating, amyloidotic form of the host protein termed the prion protein is itself the infectious agent.

Some of the TSE diseases, including Creutzfeldt-Jakob disease (CJD) in humans and scrapie in sheep and goats, have been known to medicine and agriculture for a hundred years or more. Newly emergent TSE diseases, such as bovine spongiform encephalopathy (BSE) in cattle, variant CJD (vCJD) in humans, and chronic wasting disease (CWD) in deer and elk, are responsible for recent and ongoing epidemics. There is currently no screening test sensitive enough to detect infectivity at the low levels present early in these diseases, there is no effective therapy, and the diseases are inevitably lethal. For more details on the history, properties, current research, and unique perspectives on these agents the reader is referred to a number of reviews (Aguzzi and Polymenidou, 2004; Chesebro, 2003; Collins et al., 2004; DeArmond et al., 2003; Knight and Will, 2004; Manuelidis, 2003; McKintosh et al., 2003; Rohwer, 1991; Riesner, 2004; Somerville, 2002;.)

An important principle in the safe handling of TSE agents is that of using multiple levels of control. This philosophy recognizes that no single method of containment or inactivation is perfect, and that the safest approach is to establish multiple overlapping and reinforcing layers of containment. For example, as described in greater detail below, the multiple layers utilized by our lab include limited access, complete gowning, regular general cleaning and disinfection, and specific decontamination. Our goal is that all surfaces in the TSE facility - the floor, the door knobs, the benches, the desks - are clean (i.e., agent-free) and that there are no unidentified sources of infectivity. All surfaces are disinfected on a regular schedule, gloves are changed frequently, and decontamination follows all laboratory procedures involving infectivity. One does not need protective clothing to walk safely through the TSE facility; nevertheless, we practice complete gowning at all times and treat the floor always as potentially contaminated.

Another important consideration is that all TSE agents require handling in the laboratory using TSE-specific procedures, regardless of the "risk factor" assigned to the particular agent. BSE is required (by the USDA) to be handled under Biosafety Level 3 (BSL-3) and Animal Biosafety Level 3 (ABSL-3) containment and precautions, because of its tremendous agricultural importance and potential for economic harm, and because it can cross the species barrier and infect humans as vCJD. Scrapie, which has been recognized as a disease of sheep for hundreds of years and, as far as is known, has never infected humans, is allowed to be handled under BSL-2 conditions. The classification of CWD, which is emerging as an epidemic in western and midwestern North America, is still evolving, and some institutions allow work with it in a BSL-2 laboratory while others require a BSL-3 laboratory. Yet all of these diseases, regardless of risk factor assignment, share similar properties, and to avoid contamination of the workplace, strict TSE-specific procedures must be followed.

Some of the containment measures that we follow are beyond what is required by regulatory agencies. We instituted them for an extra margin of safety for personnel and for experimental integrity. We have carried out experiments over the past twelve years in the same laboratory, with purified preparations of brain containing extremely high levels of infectivity alongside experiments with blood, in which infectivity is 10 orders of magnitude lower, and have avoided cross-contamination. We describe here how we manage the TSE ABSL-3 facility ("TSE facility") in a way that provides us with personal safety and data integrity.

Transmission

The natural means by which TSEs are transmitted outside the laboratory are not fully understood. Transmission of TSEs by aerosols has not been demonstrated. There are indications that scrapie in sheep and CWD in deer may be transmitted by casual contact, such as nose-to-nose social contact or contact with infected pastures or birth waste (Andreoletti et al., 2002; Miller and Williams, 2004; Thorgeirsdottir et al., 1999; Tuo et al., 2002;). vCJD in humans clearly comes from contact with BSE, generally presumed to be from eating contaminated tissue, but this has not been proven. We know most about TSE transmissions in nature that have resulted inadvertently from hand-of-man actions. In domestic animals, scrapie has been spread among sheep and goats by contaminated vaccines and BSE has been transmitted to cattle by feeding meat and bone meal derived from infected animal tissue (Caramelli et al., 2001; Gordon, 1946; see Robinson, 1996;). In humans kuru was transmitted by procedures involved with ritualistic cannibalism (Gajdusek, 1977), and CJD has been transmitted by various medical procedures, including corneal transplants, surgical use of dura mater, inoculation of contaminated growth hormone, and use of contaminated brain electrodes (Blattler, 2002). Most recently, vCJD has apparently been transmitted by blood transfusion from infected donors (Llewelyn et al., 2004; Peden et at., 2004).

In the laboratory, TSEs have been transmitted to laboratory animals by feeding and by virtually all parenteral routes, including intravenous, intracranial, intraperitoneal, intradermal, and subcutaneous inoculation. The various routes show different rates of transmission, with ic inoculation being the most efficient Cervenakova et al., 2004; Kimberlin and Walker, 1986). Transmission by intradermal inoculation, or scratching the surface of the skin, is a surprisingly efficient route of transmission (Taylor et al., 1996). This observation alerts workers to the need for skin protection at all times when working with the TSE agents. Nevertheless, in our laboratory at the VA Medical Center, in 12 years of experiments with over 30,000 laboratory rodents, we have encountered no evidence of horizontal transmission, either between cages of animals housed in the same room or, with the exception of transmission by cannibalism, between animals housed in the same cage. This demonstrates that it is possible to effectively manage and control TSE infectivity.

Inactivation vs. Decontamination

This chapter is primarily concerned with a practical consideration of procedures that are effective in decontaminating materials and surfaces exposed to TSE infectivity. However, an understanding of the underlying science behind these measures, and the physical underpinnings of the disinfection challenges presented by these agents, will greatly assist in their competent application. For the TSE agents, an important distinction must be made between their innate susceptibility to inactivation and their observed resistance to disinfection or decontamination. This distinction is revealed in kinetic investigations of their inactivation (Rohwer, 1984a; Rohwer, 1984b; Rohwer, 1991; Rohwer & MacAuley, unpublished). (Rohwer 1984a; Rohwer 1984b; Rohwer 1991).

TSE agents have a well-deserved reputation for being difficult to destroy. Infectivity has been recovered after 1 hour exposures to steam at 134°C (Taylor, Fernie et al.,1998), incineration at 600°C (Brown, Rau et al., 2000), 0.5% Na hypochlorite (NaClO, stock bleach diluted 1:10) (Rohwer, 1984) and 1N NaOH (Diringer & Braig, 1989; Rohwer, unpublished; Tateishi et al., 1988; Tateishi et al., 1991;) as well as after treatment with many other less severe physical and chemical disinfectants (WHO, 1999; Taylor, 2000). Many treatments that kill microbes, such as fixatives like alcohol and formaldehyde or drying onto surfaces, can actually stabilize TSE agents against some inactivants (WHO, 1999; Taylor, 2000). This resistance to disinfection has fueled endless speculation on the unconventional nature of the infectious agent and contributed importantly to serious misconceptions about its properties.

What is overlooked is that only a miniscule proportion of the total exposed TSE infectivity survives the harshest of these treatments. When the inactivations of TSE infectivity by wet heat, NaOH and NaClO have been investigated kinetically, in each case the inactivation curves were biphasic. In the initial inactivation, the vast majority of the infectivity - all but one part per million or less - was destroyed within the one or two minutes required to obtain and/or neutralize the first sample. This initial phase reveals an inherent susceptibility of TSE infectivity to inactivation by heat and alkali, that is like that of conventional microbes. The second phase of the inactivation curve exposes a refractory element, representing 1 part per million or less of the starting infectivity, that survives heat (Figure 1) or alkali for ten minutes to an hour or more. Persistence of very small, resistant subpopulation is also observed in the disinfection of conventional agents, and has been a continuing problem in killed vaccine production and water purification.

Wet Heat Inactivation at 121°C

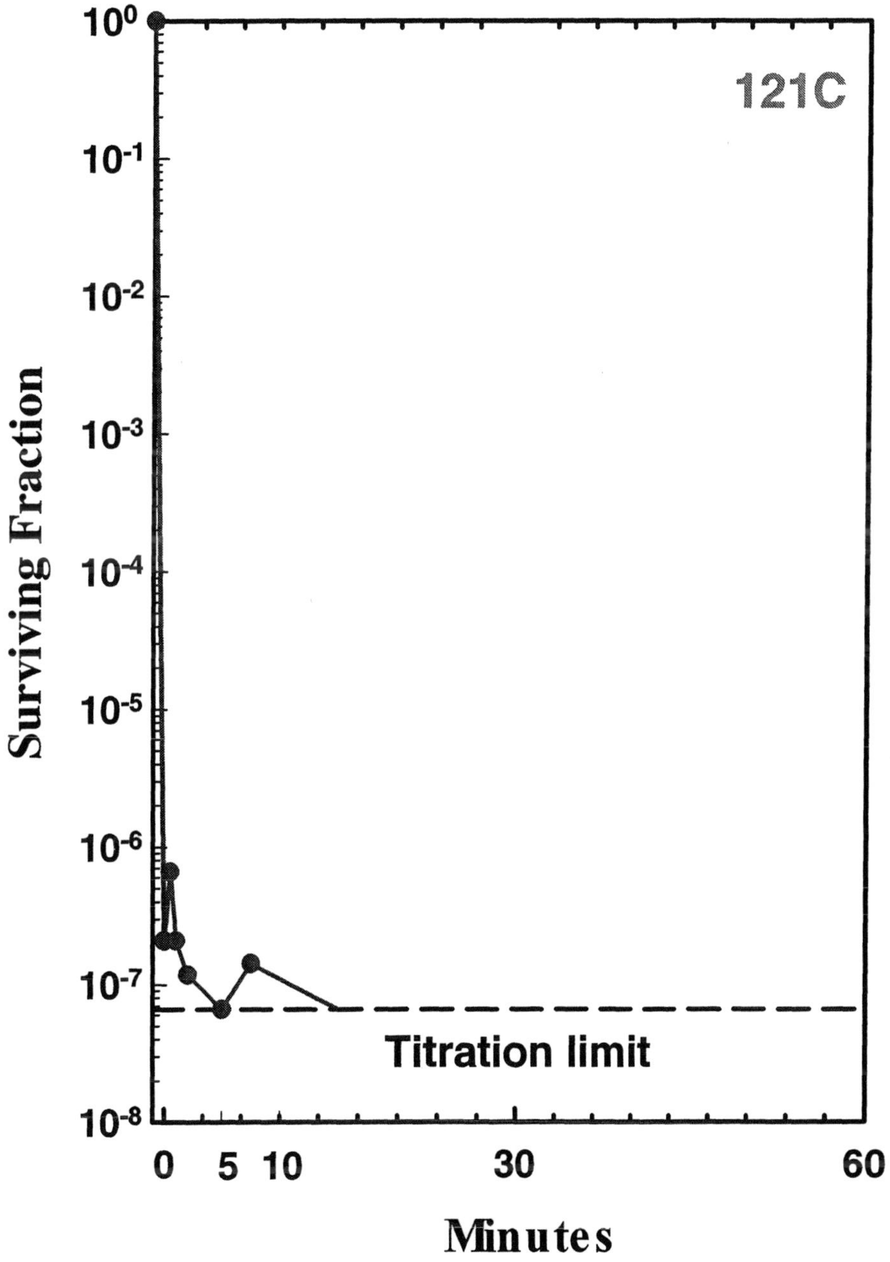

Figure 1. Inactivation curve for wet heat at 121°C

Powerful denaturants and hydrolytic treatments like wet heat and alkali will destroy any biological macromolecular structure whether a self-replicating amyloid, as proposed for TSE infectivity by the prion hypothesis, or a conventional virus or spore. Thus, whatever the structure of the TSE agent, it is unlikely that the small subpopulations that resist inactivation owe their resistance to unique intrinsic structural features of the agent. It is much more likely that they survive because, as a very rare event, at the level of a few parts per million, they never see the inactivant. How might an infectious particle escape exposure? It might never be exposed to the steam or NaOH present in the bulk environment if it were embedded in an anhydrous oil droplet or lipid raft, dried on a surface under a varnish of oxidized brain lipids or other macromolecules, or lodged in a micro pocket or crack in the surface of a container. These conditions might require far longer exposures for steam or NaOH to penetrate the protective layers around the infectivity. An infectious particle that escapes direct exposure to steam would only experience dry heat. Dry heat sterilization in general requires much higher temperatures than wet heat sterilization for the same level of effect and a one or two hour exposure to dry heat at 134°C would be of little consequence for many conventional viruses and spores as well as for TSE infectivity Brown et al., 2000; (Joslyn, 1983).

In summary, experiments on the kinetics of inactivation by NaOH and wet heat have shown that the inherent susceptibility to inactivation of TSE infectivity is within the range seen for viruses and spores, that is, most of the infectivity is nearly instantly eliminated. However, there can be measurable residual infectivity even after an hour in 134° Csteam or 1N NaOH at room temperature. This surviving infectivity, although of very low titer, is still capable of producing infections and is the source of many of the most intractable public health and animal health problems associated with the TSE agents. It is what makes the disinfection of these agents so difficult and it is behind most of the TSE-specific contamination control measures that we have implemented. However, this very unrepresentative minority fraction should not be used to draw structural conclusions about the intrinsic nature of the infectious agent itself.

Procedures used for Decontamination

The World Health Organization has produced a set of guidelines for the management of TSE infectivity (1999), to which we contributed and which we endorse. These guidelines, especially Section 6 and Annex III, contain a complete description of recommended procedures for decontamination by both chemical and thermal means that are ranked in decreasing order of stringency so as to accommodate public health settings world wide, including sites where autoclaves and incinerators are not available. Procedures higher on the list should always be used where the means to do so are available.

Below, we summarize the procedures for TSE decontamination used in our laboratory. These procedures have evolved during 25 years of working with TSE infectivity in a setting where experimental objectives have ranged from purification of high concentrations of infectivity for physical characterizations to titrations of very low infectivity tissues like blood. In some of the latter experiments, large numbers of animals inoculated with concentrations of infectivity below the limit of detection were held for two to three years without infections. We also hold at least 1% of our colony as long term uninoculated sentinels. Cross-contaminations could not be tolerated and have not been observed in these experiments.

Chemical Decontamination

Many standard biocides (e.g., formaldehyde, lysol, permangenate) are not effective or only partially effective in disinfecting TSEs (Rohwer, 1984; Taylor, 2000; WHO, 1999;). The best characterized chemical disinfectants that have been most effective for TSE disinfection are 1N (or greater) NaOH (Brown et al., 1984) and sodium hypochlorite providing at least 20,000 ppm available chlorine (prepared as a 1:2.5 dilution of 5.25% household bleach (Taylor, 1999; WHO, 1999). Our laboratory uses NaOH as the standard chemical decontaminant for instruments, equipment, and laboratory surfaces. To provide the widest possible margin of safety whenever possible, we combine chemical decontamination by NaOH with thermal decontamination by autoclaving. While there can still be measurable residual infectivity after disinfection by either method alone, this has not been the case when NaOH and wet heat have been applied either concurrently or consecutively.

We use 2N NaOH instead of 1N NaOH to provide an extra margin of safety in its application. 1N solutions of NaOH are highly effective but NaOH is slowly neutralized by reacting with CO_2 in the air and the hydroxyl radical is consumed to some extent by hydrolysis during inactivation. Also, by providing a wide margin of safety in its use we can safely tolerate the imprecision in making up stock solutions volumetrically. For this purpose we add a standard volume of dry NaOH pellets to a mark on a plastic beaker equal to 400g and transfer this to a sturdy glass bottle with a large mouth funnel. We then fill the bottle to the 1 liter mark with tap water and swirl with gloved hands to dissolve. This makes a 10 N solution which is diluted to the working stock by filling a separate 1 liter bottle to the 200 ml mark with 10 N NaOH and bringing to 1 liter with tap water. The dissolution of NaOH is highly exothermic and the solution is too hot to handle without gloved hands. NaOH 10 N and 2N stocks are present in each laboratory.

The 2N stock of NaOH will slowly take up CO_2 which will form a white precipitate over time. The 10 N stock does not take up CO_2 and is stable. The 2N stock is discarded and replaced if a heavy precipitate has formed. However, this precipitate never becomes great enough during normal use and storage to affect potency. There would have to be over 84 g/liter of carbonate precipitate to reduce the concentration of NaOH from 2N to 1N and 1N is still a highly effective concentration.

Surfaces that have been contaminated, or exposed to possible contamination, are covered with paper towels to absorb any liquid droplets. When the contamination has been absorbed, the towels are discarded and replaced with fresh towels. Sufficient 2N NaOH is poured or pipetted onto the towels to saturate them. After standing for at least 10 minutes, the towels are removed and discarded into the biohazardous trash. If there is evidence of a residue, the surface may be scrubbed with NaOH-soaked towels to further clean it. Remaining NaOH is rinsed away by repeated wiping with paper towels soaked with water. A chalky film on the surface after drying indicates that NaOH was not completely removed. The film, which may contain both residual NaOH and $NaCO_3$ formed by the reaction of NaOH with CO_2 in the air, indicates that additional wiping is required.

We use hypochlorite for decontamination only in the special circumstance where there is an incompatibility with NaOH, for example surface decontamination of aluminum rotors or buckets. At least a 5 to 10 minute exposure to undiluted bleach (~5%) is used as it is significantly more effective (than the standard household concentration of 0.5% (Rohwer, 1984; Taylor, 1999). We do not use undiluted bleach routinely because it causes more corrosion problems than NaOH and because its volatility introduces problems such as noxious fumes, especially if the clean-up waste is autoclaved.

Regular, Scheduled Cleaning of the TSE Facility

The "Environ" formulation of LpH (#6405-08, Steris Corporation, St. Louis, MO) has been demonstrated to be highly effective for sterilizing TSE infectivity in suspension (Ernst and Race, 1993; Race and Raymond, 2004). We use it in the TSE facility for regular, maintenance cleaning where no known TSE contamination has occurred, and where TSE exposure is not likely (animal rooms, general non-laboratory surfaces). Floors of the TSE facility are mopped daily with Environ LpH. Sinks, doors, and door knobs are cleaned weekly with Environ LpH.

For cleaning surfaces that are much more likely to have been exposed to TSE infectivity, we use 2N NaOH. We use NaOH so extensively because we can rely on data from our own experiments to guide us (Brown et al., 1984; Rohwer & MacAuley, unpublished), it is easy to rationalize its effectiveness, it is reasonably simple to prepare and manage, it is highly compatible with stainless steel, and most of the work surfaces in our TSE facility are stainless steel. An example of its use is the cleaning and decontamination of the work surface of the biosafety cabinet with 2N NaOH at the end of each experimental procedure. Additionally, biosafety cabinets are disassembled and the internal trays and drip pans are cleaned with 2N NaOH on a regular basis (approximately once a month).

In summary, we use LpH where regular, maintenance cleaning of large areas is involved, because LpH is easier to use than NaOH, is less caustic, leaves no residue, and is clearly adequate for removing low levels of TSE infectivity. We use NaOH for cleaning and decontaminating known spills or highly exposed areas because, while NaOH was initially adopted on the basis of an expectation of efficacy, years of experience have authenticated its effectiveness and it is now the accepted standard for chemical removal of high levels of infectivity in experimental settings.

Thermal Decontamination by Autoclaving

The USDA has specified autoclaving at 134°C for 18 minutes in a porous load autoclave for decontaminating the class 3 TSE agents BSE and vCJD. We autoclave at 134°C for 1 or 2 hours in gravity displacement autoclaves with pre-vacuum assists (Steris AMSCO Eagle Series SV3043 or 3043 Vacamatic with pre-vac upgrade). We use these longer times because we consider the time element more significant than the temperature element in thermal inactivation. The USDA standard is based on experiments using a porous load autoclave, designed for clean, dry items enclosed in porous autoclave bags or on autoclave trays. These conditions are far from the conditions in our laboratory, in which we need to decontaminate complex configurations of contaminated materials. Also porous load sterilization is inappropriate for liquids. As our kinetic experiments have shown, a steam temperature of 121°C is sufficient to inactivate TSEs, if the infectivity is actually exposed to steam at that temperature. By using longer autoclave cycle times and a pre-vacuum cycle that removes pockets of air that would otherwise displace or prevent contact with steam, our intent is to maximize infiltration of even deeply buried regions of the load by live steam which we have shown inactivates within seconds of contact. All TSE-contaminated waste bags are autoclaved for 2 hours and waste that is being autoclaved is loosely - NOT tightly-packed in open bags that are not sealed in any way. We recognize that even with these precautions, some circumstances may prevent steam access-such as inverted gloves or wadded up plastic-backed bench pads. In these cases, we rely on the fact that all waste is subsequently incinerated after autoclaving.

Routine caging and bedding, which is not expected to be contaminated since there is no evidence that infected laboratory animals normally shed infectivity, is autoclaved for 1 hour. However, caging from the first 3 changes after inoculation is autoclaved for 2 hours because of the possibility that the caging might have been contaminated from contact with the inoculation site, or due to grooming of the inoculation site and subsequent excretion in feces. Autoclaving for 2 hours gives an extra margin of safety for decontamination in these cases. We also autoclave cages for 2 hours if there has been cannibalism by cage mates, in which case the cage may become exposed to highly infectious brain tissue, directly or through feces.

In 10 years of experience with these protocols, we have seen no examples of transmission of disease to laboratory animals from contaminated caging.

Waste Ttreatment

All solid waste is autoclaved out of the facility through one of the two pre-vacuum pass through autoclaves. Each autoclave is tested daily for proper functioning of the evacuation cycle using a Daily Air Removal Test (DART) device and temperature-sensitive paper indicators to show proper temperature was reached. Weekly spore tests are used to validate sterilization effectiveness. After autoclaving, the waste is picked up by a licensed contractor and taken to a hazardous waste disposal facility where it is incinerated. The only exceptions are flammable organic solvents (see below) and animal carcasses, which are not autoclaved but are triple bagged (as described below in the Animal Procedures section) and picked up and incinerated by a biohazardous waste contractor.

Laboratory waste is placed in stainless steel pots ("kill pots") (Figure 2) lined with size-appropriate autoclavable bags. Needles, syringes, razor blades and scalpels are placed in specially designed "sharps' containers". We obtain kill pots from restaurant supply stores where they are sold as stock pots and can be obtained in sizes ranging from 1 gallon to 20 gallons. The large pots in Figure 2 are the 20-gallon size. The kill pot plus bag (or the entire sharps' container, when 3/4 full) is autoclaved as a unit at 134°C for two hours. The bag is not removed from the kill pot during autoclaving in order to prevent leakage that might result from a chance penetration of the bag by the contents of the trash, and also because the pot itself needs to be decontaminated. The only exception to this is that the bags from the kill pots near the TSE facility exit and entrance, which contain only protective clothing, are removed from the kill pot for autoclaving. To allow efficient penetration of steam to all portions of the waste during autoclaving, the waste is not tightly packed, the kill pot lid (used to cover the contents during transport to the autoclave) is removed and placed vertically next to the pot during autoclaving, the liner bag is left open and not folded over or closed in any way, and the sharps' container is left partially open. We consider wire-frame biohazard bag holders to be a biohazard in and of themselves and they are not allowed in the TSE facility.

Figure 2. Picture of kill pots covered for transport, and with lid beside it for autoclaving

Infectious liquid waste is decontaminated by adding 10 N NaOH to give a final concentration of 2N NaOH, and then autoclaved out of the facility. The only exception to this procedure is that used for flammable organic solvents- such as phenol-chloroform mixtures used to extract DNA and RNA from TSE-infected tissues. These solvents are mixed with powdered Universal Gel Sorbent (Uni-Safe by OKO-TEC, PND Corporation, Bellevue, WA) to create a gel that is then triple bagged (as with animal carcasses) and picked up with other biohazardous waste for incineration.

Incineration of Animal Carcasses

All animal carcasses are disposed of by incineration. Autoclaving is not recommended as a method of carcass decontamination even for conventional pathogens because of the impossibility of ensuring adequate penetration of inactivating steam into the interior of carcasses. Complete hydrolysis of animal carcasses using NaOH and heat, in a tissue digester specially designed for this purpose, is an alternative that may become more widely adopted, once practical, laboratory-scale digesters are validated for TSE inactivation.

As they are generated, carcasses of dead animals are placed in plastic biohazard bags (1st bag) and stored in a freezer located inside the TSE facility until the day of pick-up by the biohazardous waste contractor. On that day, the bagged carcasses are removed from the freezer and combined with other bagged carcasses into a larger plastic bag (2nd bag) while still inside the TSE facility. This bag of carcasses is taken out of the TSE facility in a two-person procedure. One person, holding the bag of carcasses, exits the TSE facility into the anteroom and stands on the dirty side of the "clean/dirty" line. A second person who is gloved but not gowned, brings a clean plastic bag to the clean side of the "clean/dirty" line and holds it open. The first person slips the bag of carcasses into the clean bag (3rd bag) being held by the second person, being careful not to touch the outside of the clean bag. The lip of the receiving bag is protected by inverting the open edge to the outside, introducing bag 2 and then bringing the folded edge closed from the outside. The second person then seals the third bag and carries it to a biohazardous bin labeled with an "Incinerate Only" sign. The bin is secured with cable ties and loaded onto a truck by the contractor and transported to a hazardous waste incinerator.

Decontamination of Instruments by a Combination of Chemical and Thermal Inactivation

In a typical hospital setting, instruments are cleaned and then decontaminated and sterilized. In our laboratory setting, we prefer to decontaminate instruments first and then clean them. However, infectivity that has dried on a surface is much more difficult to inactivate than infectivity that is dispersed in a fluid. For this reason, it is essential that surgical instruments be kept wet at all times from the time of use through the entire decontamination cycle until they are cleaned and dried. We decontaminate instruments such as dissecting scissors and forceps, by submerging them in 2N NaOH and then autoclaving at 134°C for 1 hour. Scissors are completely opened when placed in the NaOH, to assure penetration at the hinge. Only stainless steel pans are used in this procedure, never aluminum and especially never aluminum foil, as NaOH can react explosively with aluminum at the elevated temperatures in the autoclave. Sufficient NaOH is added to the pan initially to ensure that the instruments will remain fully covered even if there is evaporation or spillage. Also, the pan containing NaOH is nested inside a larger pan, so that if NaOH sloshes out during handling, it will be contained and not fall onto the floor or onto laboratory personnel.

Hot NaOH is aggressively corrosive and we use full face shields, heavy neoprene aprons, and gloves and boots while removing and handling containers filled with hot NaOH from the autoclave.

After autoclaving in NaOH, the instruments are rinsed extensively in water and then cleaned by hand and by sonication for 90 minutes in a highly alkaline detergent (Branson IS Formulated Cleaning Concentrate) at 60°C. Instruments are then air dried, packaged in autoclave pouches, and sterilized at 125°C for 30 minutes on a dry cycle.

General Procedures in the TSE Facility

Access

We work with all TSE agents, both Level 2 agents such as scrapie and CJD and Level 3 agents such as BSE and vCJD, in an ABSL-3 facility which is comprised of a suite of two animal rooms, a procedure room, and three laboratories, all behind a locked, double-door anteroom. Within the anteroom, behind the first locked door, are a change area, a shower room, and a BSL-2 laboratory and office area. Access to the TSE facility is limited to laboratory personnel whose duties require them to work there, and to authorize visitors such as inspectors, repair/maintenance personnel, or academic visitors taking a scheduled tour of the facility to learn about its operation. Laboratory personnel experienced in our TSE procedures always accompany visitors.

As a result of increased sensitivity to the potential for bioterrorism, the level of physical security for our facility has recently been increased and access is now controlled by a computerized system of electronic locks and digital keys that identify the person using the key. The system requires and our protocols stipulate that each person uses their assigned key to enter and exit, even if entering with a group. Visitors without keys sign a visitors log after entering the anteroom with approved personnel. A video camera connected to the medical center's police operations center monitors the entrance to the TSE facility.

Protective Clothing

Personnel wear fully protective clothing for all work within the TSE facility. We have used protective clothing from a number of manufacturers and find that provided by Kimberly Clark (Roswell, GA) is the most comfortable and suits our clothing needs best. Clothing includes a surgical gown with back-wrap opening, a surgical-type mask, bonnet that covers the hair and ears, booties, and double nitrile gloves. (Figure 3)

In addition to the gown, disposable long pants are available and used by personnel whose work requires contact with the floor, such as equipment repairmen and service technicians. For homogenization procedures performed in a biosafety cabinet, personnel wear disposable sleeves, from cuff to elbow, over the double nitrile gloves and gown. Sleeves, gowns and any other relevant protective clothing are changed immediately after any known contamination. Gowns are changed at the end of a large experiment.

Those individuals who do not wear eyeglasses, wear safety glasses or face shields when performing any procedure in the TSE facility. All personnel (including those who wear eyeglasses) always wear safety glasses or face shields when working with hazardous chemicals or highly infectious, purified preparations.

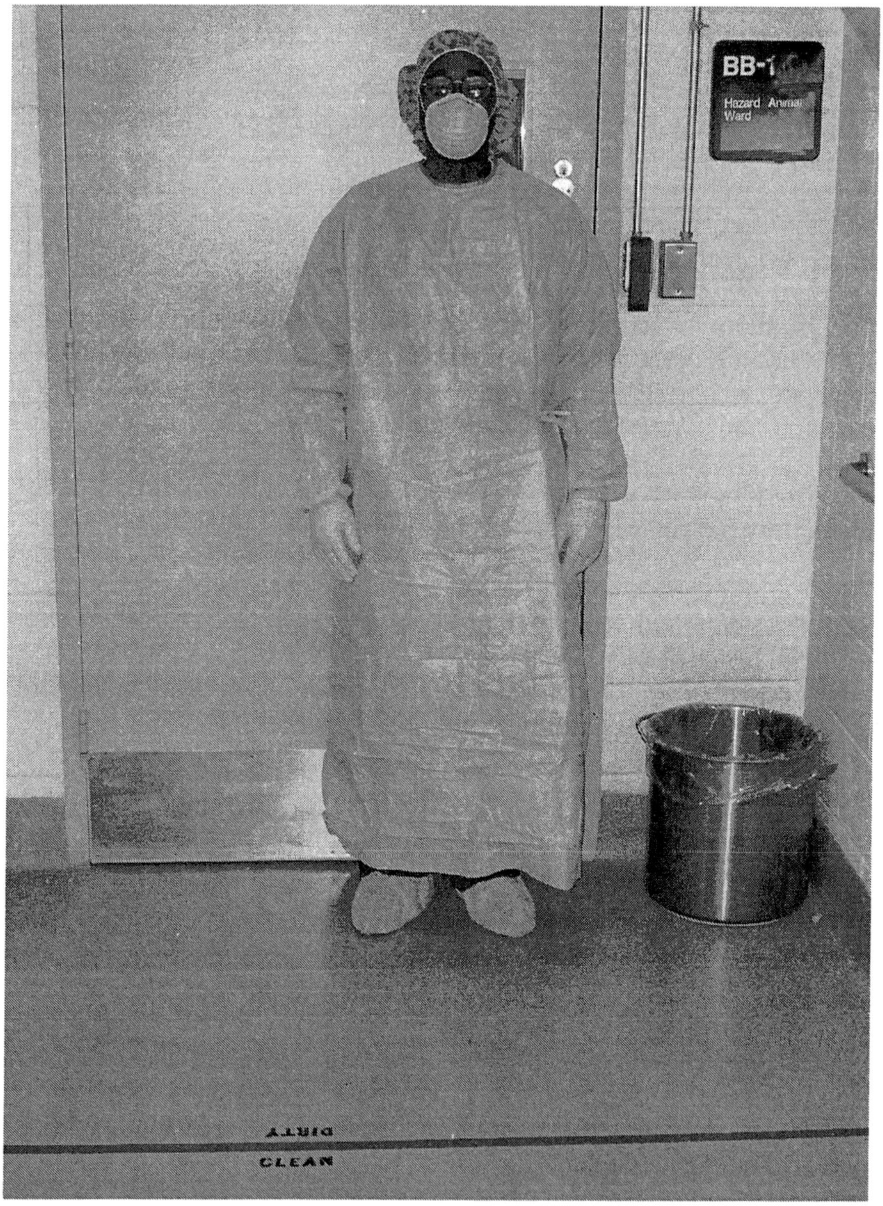

Figure 3. Person fully gowned

Latex gloves are not used in the TSE facility. Nitrile gloves have the advantage that defects in the glove become immediately apparent when the glove is initially stretched and put on. Outer gloves are changed:

1. Whenever the gloves are known to have become contaminated
2. Whenever there is even a remote possibility that they might have become contaminated
3. At least once every half hour

4. Before touching any surface (telephone, door-knob, etc.) not directly involved in an experiment
5. At other times as designated by experiment-specific protocols.

All protective clothing is discarded into the biohazardous trash when exiting the TSE facility, with the exception of safety glasses and surgical gowns used by personnel conducting administrative activities or other routine activities other than laboratory procedures involving infectivity. Glasses are hung on a personal hook in the anteroom. If the gown has not become contaminated or soiled it may be hung on a personal hook inside the TSE facility and re-used for up to one week.

Laboratory Procedures

A TSE-infected brain can contain very high concentrations of infectivity - 10^{10} infectious units/gram of brain from a 263K-scrapie-infected hamster. Thus, the mass of material associated with significant infectivity can be very small, making it exceedingly difficult to detect using any mass-based assay. This is a critical problem that plagues TSE research. The lack of a sensitive diagnostic (except for the bioassay, which takes 6 to 12 months in laboratory rodents), makes it difficult to monitor TSE contamination. Standard means of monitoring, such as wipe tests, are not available. This necessitates strong measures to prevent contaminations in the first place and proactive cleaning on the presumption of contamination to manage the risk.

Biosafety Cabinets & Other Work Surfaces

All work with high concentrations of infectivity is performed inside biosafety cabinets within the TSE facility, and only by personnel specifically trained in the handling of TSE agents. All work surfaces, including sinks, tables, and mobile carts as well as biosafety cabinets, are of polished stainless steel, which readily shows even minute amounts of contaminating material on the surface. Supplies are stored in autoclavable rat cages, with microisolator lids, kept on stainless steel wire shelves. There are no drawers or cupboards.

After each experiment, biosafety cabinets are cleaned thoroughly by wiping the exposed work surfaces with 2N NaOH, allowing it to remain for up to 20 minutes, and then wiping it away with water until a chalky white film no longer forms. If at any time a spot is detected on a stainless steel work surface, it is treated as if it were infectious and is decontaminated by wiping with 2N NaOH as above. Plastic-backed absorbent bench pads are placed on work surfaces during an experiment, and discarded immediately after every procedure.

Sonication

Aggregation is a major impediment in obtaining consistent and accurate results when working with TSE infectivity. TSE titers are underestimated when aggregates exist. For this reason, we routinely disaggregate TSE samples immediately before dilution and inoculation. Sonication is a highly effective means of disperson, with the advantage that sonication probes are easy to clean and decontaminate. This is not true of rotor-stator devices, which can produce excellent dispersions but entrap air more readily and must to be completely disassembled to clean and decontaminate. Sonication is also used for preparing homogenates for biochemical experiments.

All sonication is performed inside a biosafety cabinet. To provide maximum containment and minimal release of aerosols, the tube containing the material to be sonicated is capped during sonication. A hole is drilled in the cap using a bit that is the same diameter as the shaft of the sonication probe, and the probe shaft is inserted through this hole. (Figure 4) Polyproplylene caps can be easily drilled by hand, using a split point bit. This drilling is done outside the TSE facility. Because the hole can enlarge during sonication, the probe shaft is wrapped with paper toweling at the site where it enters the cap/sonication vessel and sealed with tape, before sonication begins.

Splattering by the probe can be reduced to visually undetectable levels by positioning the probe so that a node intersects the air interface of the fluid being sonicated. The location of nodes can be established by visually watching the air-water interface as a tube of water is moved up and down while sonicating. Tuning of the probe is done with the cap on the sonication vessel. For 15 ml conical centrifuge tubes, we typically position the tip 1 ½ cm to 2 cm below the meniscus for optimum mixing with minimum splatter or entrapment of air.

Sonication probes are decontaminated between uses by autoclaving in NaOH as described for instrument decontamination. For procedures where contamination between samples would confound the result, probes are changed between every sample. For procedures where each fraction has already been contaminated to some extent by its neighboring fractions during collection, there would be little point in changing the probe for each sample as long as the fractions are sonicated in the order in which they were collected. Instead, a single probe is simply wiped down between samples with 2N NaOH and rinsed thoroughly with PBS between sonications. We have tested for the amount of residual contamination after such cleaning by bioassay, and it is at limiting dilution or undetectable.

Figure 4. Picture of sonication tube, probe and bit

Preparation of Serial Dilutions

Whenever possible, it is best to perform experiments such that samples are handled in the order of increasing infectivity, with the lowest infectivity samples being handled first. This is not possible when preparing serial dilutions of a sample. In this case one starts with the highest titer sample, and produces successive samples in the order of decreasing infectivity. Another unavoidable liability in this procedure is the need to mix each sample extremely well before preparing the next dilution, which often produces a foam. If tubes with standard caps are used, and vortexed heavily, opening the cap inevitably results in bubbles bursting into the air above the tube. If the sample titer is 10^{10} infectious units (IU) per ml, even a microdrop can carry a lot of infectivity and can contaminate other tubes within a biosafety cabinet.

We have developed a system for preparing dilutions of infectious material using syringes and vials with rubber septum caps, to keep the material maximally contained at all times. The starting sample is placed in a vial and sealed with a rubber septum cap. An aliquot is removed using a needle and syringe.(Figure 5A), taking care to equilibrate pressure within the vial by injecting an equal volume of air before removing the sample. To prevent release of aerosols when the needle is withdrawn from a septum cap, and to protect the fingers while guiding the needle out of the cap, the cap-needle junction is wrapped with sterile gauze (2"x2" squares) while the needle is withdrawn. Droplet formation at the top of the septum is especially likely if there is any disequilibrium between the vial and atmospheric pressure but can also occur at the tip of the needle without any apparent provocation. To obtain an accurate volume the plunger must be immobilized as the needle is withdrawn. If unsecured the plunger may move slightly on its own to either discharge or pick up extra volume if the bottle is not at equilibrium with atmospheric pressure. When delivering the syringe contents to the next vial in a dilution series it is equally important to immobilize the plunger after discharge to prevent inadvertent discharge of the hub and needle contents into the receiving vial. A small test tube rack is used to hold syringe caps and syringes in a manner that allows hands-free capping (Figure 5B).

While using syringes does introduce the hazard of accidental needle stick, this is compensated for by the greater control over aerosol contamination and cross-contamination during the preparation of serial dilutions. Nevertheless, recognizing this risk, only persons that have practiced extensively with non-infectious samples and have demonstrated a high level of competence in handling syringes and infectivity are eligible to be trained and qualified in this procedure.

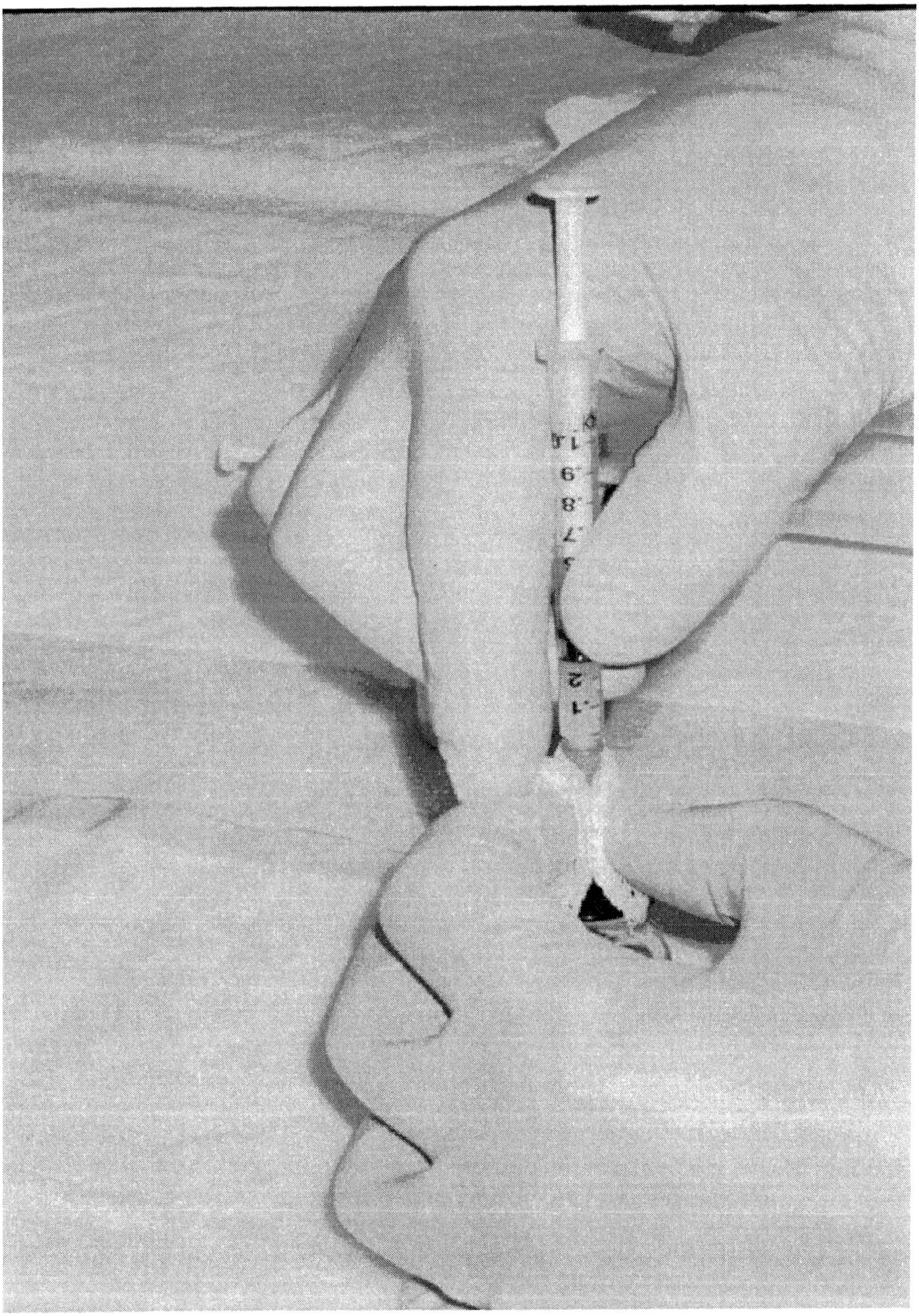

Figure 5A. Picture of needle in serum bottle, wrapped with gauze, showing way it is held

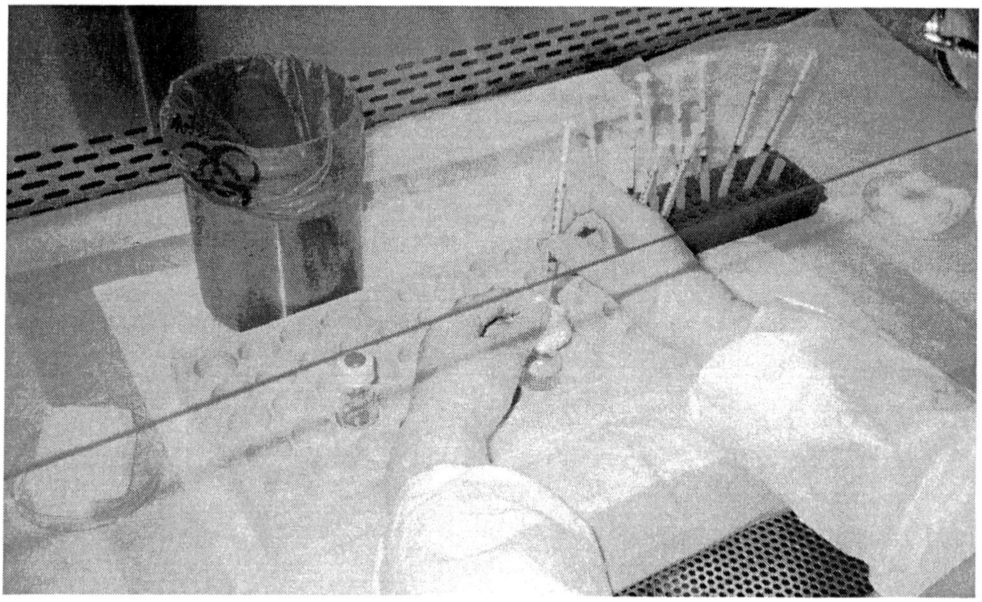

Figure 5B. Picture of Biosafety cabinet set up for serial dilutions

Serum bottles are weighed before and after every addition of mass, to provide an independent measurement and verification of the transfer of volumes and masses. We do this because the infectivity data are so costly to generate that a preventable error discovered 6 to 18 months later is inexcusable.

Bagging Equipment

Generally, all small to medium sized equipment used in TSE experiments (such as vortex mixers, columns, pumps, but not centrifuges or freezers) is bagged before use. (Figure 6A) The equipment is either placed inside a plastic bag and sealed with tape or wrapped with plastic sheeting and taped. After bagging, the equipment is placed inside a biosafety cabinet while being used with infectivity. Other equipment in the TSE facility, such as periodically replaced bottled gas tanks, is bagged with plastic before being taken into the TSE facility. The plastic is removed prior to leaving the TSE facility.

New procedures that involve working with customized equipment are subjected to dry runs with non-infectious material, to determine the potential sources of contamination and to permit development of specific procedures that are least likely to lead to contamination. Resin columns or filtration apparatuses are covered at the joints with plastic sleeves, to contain any leaks if they were to occur. We use plastic sheet tubing (cylindrical plastic bagging material) in rolls of 1,000 feet and in three different diameters (Plastic Bag Company, Inc., Hudson, OH). This material is cut to an appropriate length for plastic sleeving, and taped in place around the joint. (Figure 6B) In the event of a leak around a joint, the

TSE-exposed material drips into the sleeve, is detected visually, and the system is shut down. This principle can be extended to the entire apparatus when the possibility of leaks is considered high.

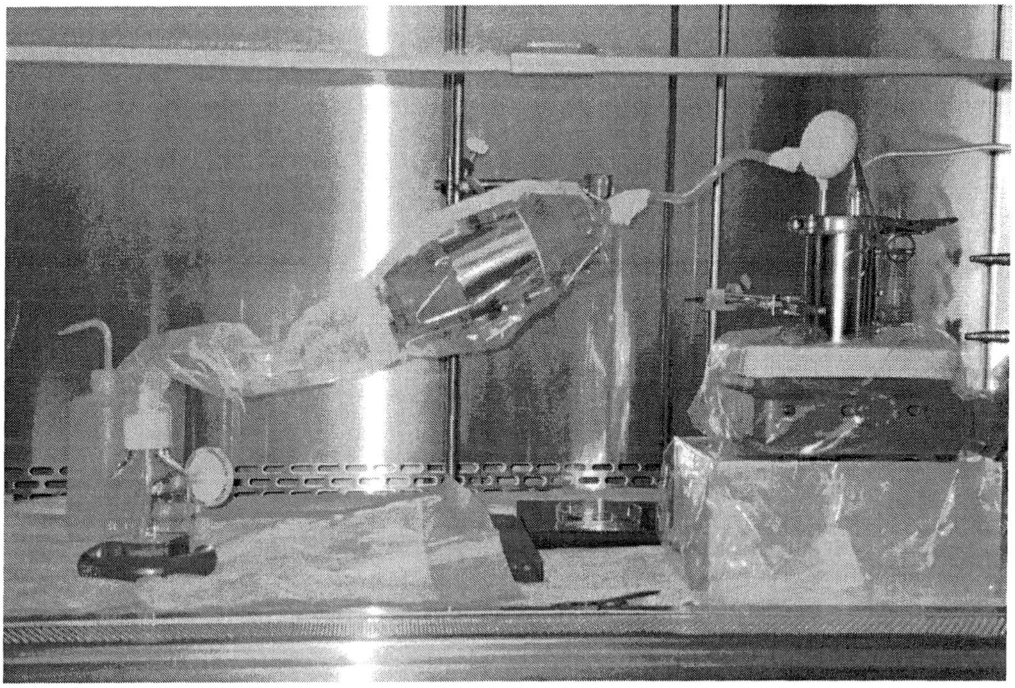

Figure 6A. Picture of small bagged equipment

Some equipment may not fit into a biosafety cabinet (such as long chromatography columns). In these cases, a containment bag can be constructed using 4 ml plastic sheets from large plastic bags or drop cloths purchased at a hardware store. The equipment supports are used as framing or a frame is built from stainless steel rods, and the entire assembly is placed on the plastic sheeting. In addition, plastic sleeves can be used on connections and tubing, as described above. To allow ease of manipulation access slits can be made in the sleeves and gloves can be taped into the sleeves. For added protection, paper towels can be placed in the bottom and saturated with 2N NaOH. In the event that a leak occurs, the NaOH can effectively inactivate the gross spill without the need to open the plastic enclosure. At the end of the experiment, the plastic and paper towel absorbent are placed in a bag-lined kill pot and autoclaved.

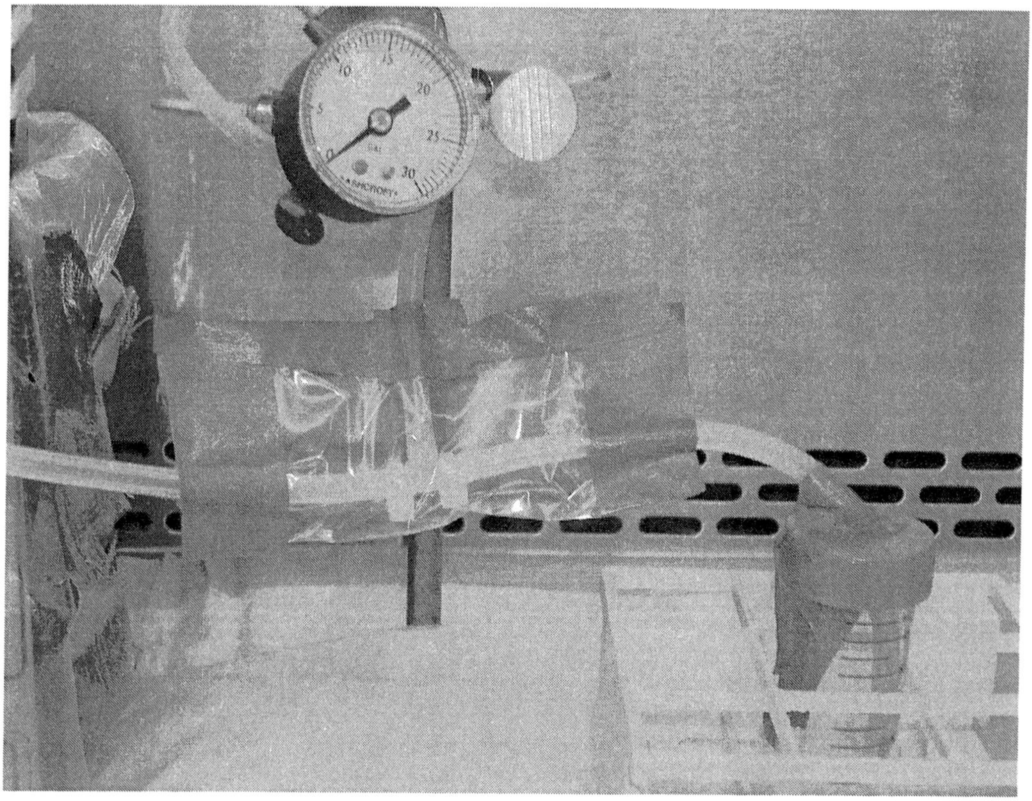

Figure 6B. Equipment with sleeves wrapped around joints]

Animal Procedures

Animal Husbandry

All TSE-infected animals are housed within our dedicated TSE facility and cared for using ABSL-3 protocols, including animals infected with scrapie. The scrapie-infected and BSE-infected laboratory rodents (hamsters, mice and rats) are maintained in microisolator caging, which is rodent caging covered with "bonnets" lined with filter paper. Studies looking at the infectivity of urine and feces have been somewhat contradictory, and we are currently investigating this further. In our facility, when laboratory animals are inoculated with a low dilution of TSE infectivity, only one animal in a cage may develop disease, and its cage mates may remain healthy for up to their normal lifespan. Thus, it appears that the cage environment of these animals is not contagious.

Immediately after any type of parenteral inoculation of TSE infectivity with a needle, the inoculation site is unavoidably contaminated. Sometimes, the wound will ooze, and allow infectious inoculum to escape. Since the animals groom themselves continuously, there is potential for the cage and bedding to become contaminated immediately after inoculation with a TSE inoculum. In addition, following oral inoculation with high concentrations of TSE infectivity, we consider it likely that the feces are contaminated. For this reason, we change all components of the caging (including bedding, bonnets, wire bar lids, water bottles, etc.) one day after inoculation and for the next two weekly cage changes. In addition, we increase the autoclave time from 1 hour to 2 hours for decontaminating this caging.

All caging, including bedding, is autoclaved at 134°C for 1 hour (or 2 hours as noted above) before emptying the bedding into the biohazardous trash and washing the caging. Cages must not be tightly stacked in the autoclave and must be separated enough to allow steam penetration. We have fabricated a frame system of stainless steel bars from which cages are hung. (Figure 7) The system holds cages apart but still allows partial nesting. This system ensures that cage decontamination is effective and consistent for each autoclaving cycle, and also prevents the cages at the bottom from carrying the weight of the caging at the top, which, with lengthy autoclaving times, can distort and misshape the caging.

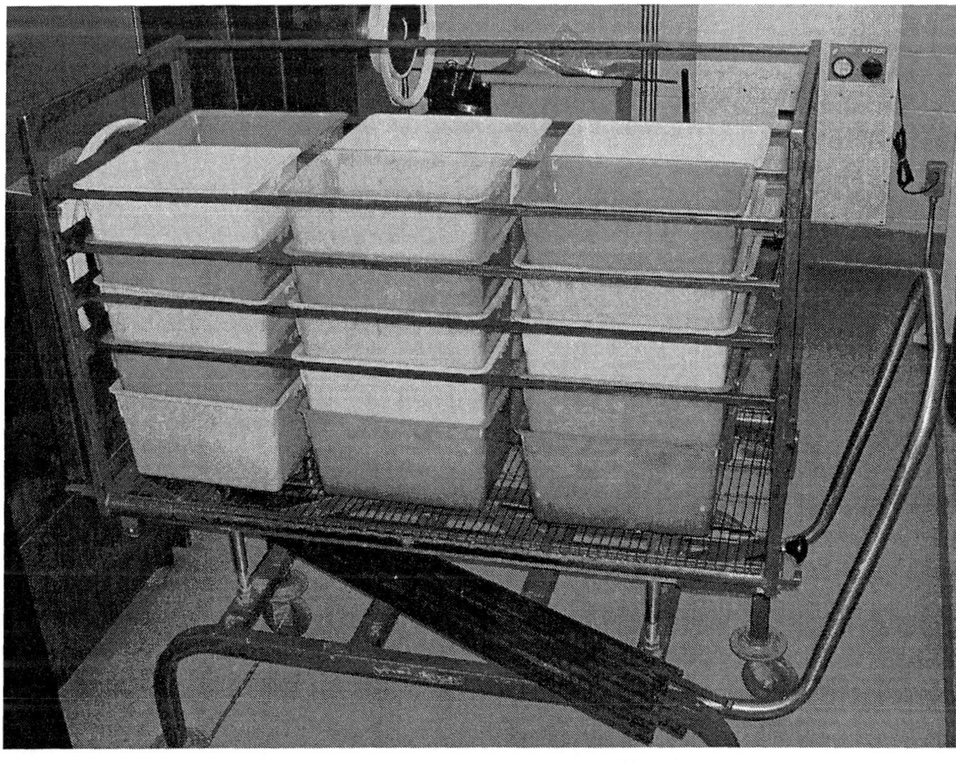

Figure 7. Caging hung from steel bar-rack

Inoculations

Because inoculations unavoidably require the use of syringes filled with highly concentrated infectivity, these procedures require carefully thought out protocols and meticulous training. To prevent accidental needle sticks, the inoculator maintains tight control over the syringe needle and limits the range that the needle can move during an inoculation by holding the syringe between the thumb, index and middle finger of one hand while stabilizing that hand against the other hand and inoculation surface. The other hand must simultaneously immobilize the animal and position it to receive the injection. This skill is transferred by one-on-one tutoring. Extensive practice with uninfected inoculum is required to obtain the confidence and proficiency to conduct the procedure with infected materials. Additional support is in the form of detailed, written protocols governing filling syringes with TSE-infected material and for intracranial, intraperitoneal, and intravenous inoculations with scrapie and BSE. Scrapie inoculations are performed in the animal procedure room in the TSE facility; BSE inoculations are performed in a biosafety cabinet.

Whenever possible, inocula are transferred to septum-topped vials for injection. This provides greater control of the infectivity and better protection against cross-contamination in the inoculation theater. With care the only contaminated surface will be the outside wall of the needles where they have penetrated the septum vials to retrieve the inocula. This is far preferable to working with open tubes from which aerosols may be released when the caps are removed and a good portion of the syringe barrel may be contaminated during retrieval of the sample. Since we do not have to uncap the samples, it also allows us to vigorously vortex each inoculum immediately before withdrawal from the vial without concern about foams, aerosols and attendant contamination. The inocula are withdrawn following the same procedures described for volume transfers during serial dilutions. The principle source of potential contamination during serial dilutions, working from the most concentrated material to the least concentrated, is absent during the inoculation of the same samples where we can work in the reverse order, from the most dilute to the most concentrated material. Since the inoculation of each titration begins anew with the least concentrated sample, the inoculation field is renewed with new bench pads and other materials between samples.

Bleeding

We have shown that the blood of TSE-infected animals contains infectivity at very low levels (Elliott et al., in preparation; Gregori et al., 2004). In hamsters with scrapie, the level of infectivity in blood is about 10 IU/ml, much less than that of nervous tissue (10^{10} IU/g in brain). Thus, in experiments involving blood, extreme measures must be taken to avoid contaminating the blood with any nerve-derived infectivity. In most of our experiments, bleeding is a terminal procedure, performed with the animal under deep anesthesia. We bleed by cardiac puncture of the right ventricle, after the diaphragm and pericardium have been resected. We take extreme care that the syringe needle does not contact any tissue other than the heart. If an animal is to be bled and the brain is also to be harvested, we bleed first by cardiac puncture, and then, after the blood has been removed from the room and stored in the refrigerator or freezer, harvest the brain. Scrapie-infected animals are bled in the procedure room; BSE-infected animals are bled in a biosafety cabinet.

Dissections

Dissections of scrapie-infected animals are done in the procedure room; dissections of BSE-infected animals are done in a biosafety cabinet. If organs are being pooled for preparation of a stock material, a single set of instruments can be used for collection, rinsing as necessary in water between uses. If organs are being harvested for later bioassay of individual animals, then each dissection is done with separate sterile instruments.

Training of Personnel

Well-trained personnel are indispensable to the successful management of TSE infectivity. A visceral appreciation that TSE agents are not visible, not easily detectable, and not easy to kill must be inculcated at all levels, from investigators to laboratory technicians to animal caregivers and technicians. We have developed an SOP-based training program in which all laboratory personnel are personally trained by an experienced, authorized person in the laboratory. Each procedure has a designated trainer. Not until an individual has been approved by this trainer, are they authorized to carry out procedures with TSE infectivity. The laboratory keeps a master list of all laboratory personnel and the procedures they have been authorized to conduct. In addition to the initial training, we conduct regular audits of performance of procedures and regular SOP review sessions as part of our laboratory meetings. Each review begins and ends with a short quiz that documents the level of individual understanding of the content. A training notebook is used to keep track of individual training records and review sessions.

Permits and Inspections

Specific permits from the USDA are required for transporting and working with scrapie, BSE, vCJD, and CWD. The permit specifies the infection control procedures to be followed with the particular agent covered. The USDA also performs an initial inspection of any BSL-3 facility requesting permission to work with BSE, and conducts annual "compliance" inspections of these facilities. A permit from the CDC is also required for transporting BSE and vCJD.

In the implementation of the current system of precautions against bioterrorism, BSE was listed as a "Select Agent," while its human counterpart, variant

Creutzfeldt-Jakob disease, v-CJD, was not. All current evidence suggests that these two agent strains are identical. The discrepancy in classification results from the designation of BSE as an "agent of high consequence to agriculture" by the USDA and thereby subject to exploitation by terrorists. In contrast the risk from vCJD was not considered by the CDC to be significantly different from that posed by classical forms of the disease, nor sufficiently attractive as an instrument of terror to be listed as a select agent. The select agent classification means that all laboratories working with this agent must register with the USDA.

Shipping

Shipping of any of these agents requires that the person who prepares the materials for shipment and signs the air waybill must be certified to ship Dangerous Goods. Certification involves taking an authorized course in Dangerous Goods shipping and requires a refresher course at least every two years. TSE infectivity can only be shipped to recipients possessing a current permit to receive and possess the agent. Our laboratory requests and keeps on file a copy of the permit and shipping documentation of every laboratory that receives infectious materials from us. As long as the permit is current, it does not have to be re-sent for each shipment.

Regulations of the International Air Transport Association (IATA) require that Dangerous Goods shipments be labeled with a phone number that will be answered 24 hours a day, 7 days a week, by a person, not a recording. That person must be able to provide background information on the contents of the shipment and containment and decontamination advice if the agent is released. We utilize a 24-hour emergency response service (Chem-Tel, Tampa, FL) to cover this requirement for us. We provide them with information for decontamination. Were there to be an accident, the service would give direction to first responders and then contact the laboratory for further instructions.

Quality Assurance and Administration

In addition to the personnel performing the actual laboratory and animal procedures, our laboratory has identified a number of other management roles that are necessary for trouble-free execution of animal-based TSE studies. Persons in these management roles have responsibility for the myriad administrative tasks that would otherwise completely overwhelm the scientific program. They interact with personnel at the bench and the institutional BioSafety Committee and regulatory bodies such as the USDA.

Internal Biosafety Officer

The primary protections for working with the TSE agents are procedural. Protocols must be carefully thought out and documented in writing, and personnel must be trained. An important link in this "protection by procedures" is assurance that the procedures are being followed. To this end, our laboratory has appointed an internal biosafety officer, who regularly and spontaneously audits procedures being performed within the TSE facility. Personnel who are not performing a procedure correctly are cited in a written report (that is reviewed by the Laboratory Director) and retrained. The biosafety officer has the authority to revoke TSE facility privileges of investigators or technicians who do not carry out procedures correctly.

In addition to this auditing responsibility, our biosafety officer, as the senior animal technician with the most TSE experience, is also responsible for most of the TSE-specific safety training in the laboratory. This is accomplished both through one-on-one technical training, and through monthly seminar-based training of the entire laboratory group in standard operating procedures. Our internal biosafety officer also trains first responders, including the police force at our own institution, Baltimore City firemen, emergency medical crews, and Hazmat personnel. All training is documented in a training notebook.

Manager of the TSE Facility

This individual supervises all animal husbandry and technical work, assigns personnel to tasks in the TSE facility, enforces security procedures in the facility, interfaces with the investigators in animal-based studies, assures that the maintenance and cleaning schedules are followed, orders and keeps inventory on animals, ensures that IACUC protocols are in place for every animal ordered and every procedure performed, and that the TSE facility is functioning properly.

Freezer Manager

The freezer manager supervises the storage and retrieval of thousands of freezer samples in six different -80°C freezers on-site and at a repository. Responsibilities include auditing the Freezer database, to ensure that laboratory personnel are properly entering identifying information for samples that are stored in the freezers, and auditing the actual freezer storage, to ensure that the location of all samples is instantly and accurately available. This function has assumed added significance with the increasing requirements for accountability of materials with bioterrorist potential.

Animal Data Manager

To capture clinical data accurately, and to interpret and make use of it, we track each animal that is received by the facility and enter information about the use of the animal (dates of birth, receipt, use, death; which study, what inoculum, results, etc.) into a database. We also track all animal studies by compiling clinical scores and creating easy-to-read charts for investigators. Our TSE studies typically involve large numbers of animals that are held for long periods of time. The animals must be monitored for disease symptoms over this entire period. This generates huge amounts of clinical data. This data is transferred out of the TSE facility by scanning to a local computer network using a dedicated document scanner or, for quick transfers of small amounts of data, by faxing to machines outside the TSE facility. Scanned data is archived onto CDROMs and original paperwork is removed from the TSE facility after studies are completed. (All original papers generated in the TSE facility are double bagged before removal from the facility and stored in a locked room as biohazardous materials, to be opened only if returned to the TSE facility. After a designated holding time, they are incinerated as biohazardous trash.) The animal data manager performs these data management tasks.

Laboratory Manager

In addition to other duties not necessarily related to the TSE facility, this person is responsible for supervising and coordinating the efforts of the above managerial personnel, maintaining up-to-date and accurate training records for all laboratory personnel, and developing and maintaining up-to-date written SOPs for all relevant procedures.

Special Problems

Along with the special procedures for handling TSE infectivity come special problems.

Most standard animal caging does not withstand repeated autoclaving for 1 to 2 hours at 134°C. Polycarbonate caging, for example, is destroyed within weeks. In contrast our polypropylene caging has lasted for over a decade under these conditions. Caging made from the new materials RaTEMP (Allentown Caging Equipment Co., Inc., Allentown, NJ) and Zytemp (Lab Products, Inc., Seaford, DE) have now also withstood this treatment for a number of years. These materials have the advantage that, unlike polypropylene, they are clear.

Long autoclaving times at high temperatures tend to bake bedding onto the cages. The latest generation of automated cagewashers with very high-powered jets can remove this residue. Otherwise, hand scrubbing may be required to get the cages completely clean.

The harsh decontamination procedures used in the lab give rise to higher than normal attrition rates for dissecting instruments, sonication probes and other laboratory equipment. This results in a fairly high replacement rate for these items. Some laboratory supply plastics (disposable petri dishes, plastic bottles, test tube racks) break down with long autoclaving at high temperatures and emit noxious fumes. This problem can be reduced by immersing plastic items in water during autoclaving.

The prolonged autoclave time and elevated temperature of autoclave cycles results in the generation of considerable environmental heat. This heat can create significant problems in an enclosed BSL-3 facility, both for personnel in protective clothing and for heat and humidity-sensitive equipment such as computers. This problem is mediated by vents over the autoclave area that vent to the outside through HEPA filters, enhanced air-conditioning, and independent thermostat control of the autoclave area.

HEPA filtration of exhaust air is not required for regular BSL-3 facilities. It is a USDA requirement for working with BSE. There is no evidence that BSE, or any TSE, is an air-borne pathogen. Our facility was retrofitted with HEPA-filtered exhaust at considerable expense to meet this requirement. Ideally, there would be a separate BSL classification and set of requirements for BSE and other TSE agents that was tailored to the specific requirements for the safe investigation of these particular agents.

It is expensive to mount a TSE research program. The procedures required for working safely with infectious TSE agents are labor and resource intensive. The protective clothing, HEPA filtration, waste processing, cage processing, attrition of equipment and materials, animal monitoring, data collection, sample storage and inventory management during the long periods between experiments and results, maintenance of high-use equipment like autoclaves, cage wash, and freezers, heating and cooling single pass air, security, and training are all costly but necessary for a successful program.

Finally, procedures take longer when they must be carried out in a containment facility and this limitation is exacerbated by the slow turn around times of TSE infectivity experiments. It is essential to keep the facility operating smoothly at a high level of productivity to justify its expense and existence.

Summary

TSEs or prion disease agents are difficult to disinfect and difficult to monitor for contamination. Infected brain tissue can contain very high titers of infectivity with a poorly defined risk to humans. TSE agents can be handled safely with strict adherence to procedures for decontamination and containment which are specific to these agents and are in general much harsher than those required for more conventional pathogens.

Acknowledgments

The authors thank Dr. Richard Gilpin for helpful comments on the manuscript and Ms. Johari Barnes for help with the figures and bibliography. The work in our lab is supported by a VA Merit Review to RGR, by NIH contract N01-NS-0-2327, by NHLBI grant R01-HL63930 and by DOD NPRP grants DAMD 17-03-1-0746, DAMD 17-03-1-0756, DAMD 17-03-1-0749.

References

Aguzzi, A., & Polymenidou, M. (2004). Mammalian prion biology: one century of evolving concepts. Cell, **116**(2), 313-327.

Andreoletti, O., Lacroux, C., Chabert, A., Monnereau, L., Tabouret, G., Lantier, F., Berthon, P., Eychenne, F., Lafond-Benestad, S., Elsen, J.M., & Schelcher, F. (2002). PrP(Sc) accumulation in placentas of ewes exposed to natural scrapie: influence of foetal PrP genotype and effect on ewe-to-lamb transmission. Journal of General Virology, **83**, 2607-2616.

Blattler, T. (2002). Implications of prion diseases for neurosurgery. Neurosurgery Review, **25**(4), 195-203.

Brown, P., Rau, E.H., Johnson, B.K., Bacote, A.E., Gibbs, C.J., Jr., & Gajdusek, D.C. (2000). New studies on the heat resistance of hamster-adapted scrapie agent: threshold survival after ashing at 600 degrees C suggests an inorganic template of replication. Proceedings of the National Academy of Science, **97**(7), 3418-3421.

Brown, P., Rohwer, R.G., & Gajdusek, D.C. (1984). Sodium hydroxide decontamination of Cretuzfeldt-Jakob disease virus. New England Journal of Medicine, **310**, 727.

Caramelli, M., Ru, G., Casalone, C., Bozzetta, E., Acutis, P.L., Calella. A., & Forloni, G. (2001). Evidence for the transmission of scrapie to sheep and goats from a vaccine against Mycoplasma agalactiae. Veterinary Record, **148**(17), 531-536.

Cervenakova, L., Yakovleva, O., McKenzie, C., Kolchinsky, S., McShane, L., Drohan, W.N., & Brown, P. (2003). Similar levels of infectivity in the blood of mice infected with human-derived vCJD and GSS strains of transmissible spongiform encephalopathy. Transfusion, **43**, 1687-1694.

Chesebro, B. (2003). Introduction to the transmissible spongiform Encephalopathies or prion diseases. British Medical Bulletin, **66**, 1-20.

Collins, S.J., Lawson, V.A., & Masters, C.L. (2004). Transmissible spongiform encephalopathies. Lancet, **363**(9402), 51-61.

DeArmond, S.J., & Prusiner, S.B. (2003). Perspectives on prion biology, prion disease pathogenesis, and pharmacologic approaches to treatment. Clinlcal Laboratory Medicine, **23**(1), 1-41.

Diringer, H., & Braig, H.R. (1089). Infectivity of unconventional viruses in dura mater. Lancet, **1**(8635), 439-440.

Ernst, D.R., & Race, R.E. (1993). Comparative analysis of scrapie agent inactivation methods. Journal of Virological Methods, **41**, 93-202.

Gajdusek, D.C. (1977). Unconventional viruses and the origin and disappearance of kuru. Science, **197**(4307), 943-960.

Gordon, W.S. (1946). Advances in veterinary research. Veterinary Record. **58**, 516-520.

Joslyn, L. (1983). Sterilization by heat, pp.3-46. In Seymour Block (Ed.), Disinfection, Sterilization, and Preservation, Philadelphia, PA: Lea and Febiger.

Kimberlin, R.H., & Walker, C.A. (1986). Pathogenesis of scrapie (strain 263K) in hamsters infected intracerebrally, introperitoneally or intraocularly. Journal of General Virology, **67**(2), 255-263.

Knight, R.S.G., & Will, R.G. (2004). Prion diseases. Journal of Neurology, Neurosurgery, and Psychology, **75**(Supple 1):136-142.

Llewelyn, C.A., Hewitt, P.E., Knight, R.S.G., Amar, K., Cousens, S., MacKenzie, J., & Will, R.G. (2004). Possible transmission of variant Cretuzfeldt-Jakob disease by blood transfusion. Lancet, **363**, 417-421.

Manuelidis, L. (2003). Transmissible encephalopathies: speculations and realities. Viral Immunology, **16**(2), 123-139.

McKintosh, E., Tabrizi, S.J., & Collinge, J. (2003). Prion diseases. Journal of Neurovirology, **9**(2), 183-193.

Miller, M.W. & Williams, E.S. (2004). Chronic wasting disease of cervids. Current Topics in Microbiology & Immunolology 284:193-214.

Peden, A.H., Head, M.W., Ritchie, D.L., Bell, J.E., & Ironside, J.W. (2004). Preclinical vCJD after blood transfusion in a PRNP codon 129 heterozygous patient. Lancet, **364**(9433), 477-479.

Race, R.E., & Raymond, GJ. (2004). Inactivation of transmissible spongiform encephalopathy agents by Environ LpH. Journal of Virology, **78**(4), 2164-2165.

Riesner, D. (2004). Transmissible spongiform Encephalopathies: The prion theory—Background and basic information. Contributions to Microbiology, **11**, 1-13.

Robinson, M.M. (1996). Transmissible Encephalopathies and biopharmaceutical production. Developmental Biological Standards, **88**, 237-241.

Rohwer, R.G. (1984a). Scrapie infectious agent is virus-like in size and susceptibility to inactivation. Nature, **308**, 658-668.

Rohwer, R.G. (1984b). Scrapie shows a virus-like sensitivity to heat inactivation. Science, **223**, 600-602.

Rohwer, R.G. (1991). The scrapie agent—A virus by any other name. Current Topics in Microbiology & Immunology, **172**, 195-232.

Somerville, R.A. (2002). TSE agent strains and PrP:reconciling structure and function. Trends in Biochemical Science, **27**(12), 606-612.

Tateishi, J., Tashima, T., & Kitamoto, T. (1988). Inactivation of the Creutzfeldt-Jakob disease agent. Annals of Neurology, **24**(3), 466.

Tateishi, J., Tashima, T., & Kitamoto, T. (1991). Practical methods for chemical inactivation of Creutzfeldt-Jakob disease pathogen. Microbiology & Immunology, **35**(2), 163-166.

Taylor, D.M. (1999). Inactivation of prions by physical and chemical means. Journal of Hospital Infections, **43**(Suppl), S69-76.

Taylor, D.M. (2000). Inactivation of transmissible degenerative encephalopathy agents: A review. Veterinary Journal, **159**(1), 10-17.

Taylor, D.M., Fernie, K., McConnell, I., & Steele, P.J. (1998). Observations on thermostable subpopulations of the unconventional agents that cause transmissible degenerative encephalopathies. Veterinary Microbiology, **64**(1), 33-38.

Taylor, D.M., McConnell, I., & Fraser, H.(1996). Scrapie infection can be established readily through skin scarification in immunocompetent but not immunodeficient mice. Journal of General Virology, **77**(7), 1595-1599.

Thorgeirsdottir, S., Sigurdarso,n S., Thorisson, H.M., Georgsson, G., & , Palsdottir, A. (1999). PrP gene polymorphism and natural scrapie in Icelandic sheep. Journal General Virology, **80**(9), 2527-2534.

Tuo, W., O'Rourke, K.I., Zhuang, D., Cheevers, W.P., Spraker, T.R., &, Knowles, D.P. (2002). Pregnancy status and fetal prion genetics determine PrPSc accumulation in placentomes of scrapie-infected sheep. Proceedings of the National Academy of Sciences, USA, **99**(9), 6310-6315.

WHO (World Health Organization). WHO infection control guidelines for transmissible spongiform Encephalopathies. Report of a WHO consultation. Geneva, Switzerland, World Health Organization, 1999.